Machine Learning and Non-volatile Memories

“MEDUSA”, polychrome bronze, Rome 2021, by Alessandro Romano (www.alessandroromano.com). Photograph by Valerio Ventura

Rino Micheloni · Cristian Zambelli
Editors

Machine Learning and Non-volatile Memories

Springer

Editors
Rino Micheloni
Engineering Department
Università degli Studi di Ferrara
Ferrara, Italy

Cristian Zambelli
Engineering Department
Università degli Studi di Ferrara
Ferrara, Italy

ISBN 978-3-031-03840-2 ISBN 978-3-031-03841-9 (eBook)
https://doi.org/10.1007/978-3-031-03841-9

This Springer imprint is published by the registered company Springer Nature Switzerland AG
The registered company address is: Gewerbestrasse 11, 6330 Cham, Switzerland

To my beloved wife Sabrina, my daughters Laura and Greta, and my dog Leonidas, who make my life worth living
Rino

To my dear wife Margherita and my parents Marzia and Marco for their unconditional support and love
Cristian

Foreword

Rino Micheloni is a prolific author and editor of books and papers on the latest storage-related technologies. With this new book, *Machine Learning and Non-volatile Memories*, Rino and Dr. Cristian Zambelli (University of Ferrara, Italy) continue the success in identifying the most important developing technologies, explaining the state-of-the-art clearly, and pointing the direction for the next big advances. Artificial intelligence and machine learning (AI/ML) are still in the early stages of practicality. However, it seems clear that adding AI/ML capabilities close to where the data are generated or stored will yield the most efficient solutions.

The breadth of applications covered in this book reflects AI/ML's adoption in practically all fields. Similar to PCs, the Internet, or cloud computing, it seems evident that it is not a question of which fields will embrace AI/ML, but which will do it most quickly and extract the biggest benefit. As AI/ML is used in more products and services, the industry will converge on the most-effective, efficient architectures. This is typical in technology development lifecycles in which all practitioners see the same physics and face the same economic realities. This book enables researchers and designers to more quickly arrive at efficient, effective storage solutions, accelerating the pace of advancement of this important technology.

My company, Channel Science, is building-in machine learning to technology we are currently developing for a US Department of Energy project. We are recovering irreplaceable scientific data from deteriorating 1960s and 70s magnetic tapes. We are using ML as an integral part of reading signal waveforms to identify signal-quality issues with the tapes, apply the proper mitigation techniques, and determine the type of data being recovered. Our experience shows that machines that have context for the streams of bits and bytes they are sensing are able to do more with them and do it better.

This new book explains how designers and architects are making AI/ML computations more efficient. The authors look closely at the matrix and vector multiplications at the core of the neural networks running AI/ML algorithms, both for training and inference. These calculations are not well suited for general purpose CPUs (central processing units), but they have been more efficiently implemented on GPUs (graphics processing units). However, the growing usefulness of very large models means some high-value applications require the training of millions of parameters, and eventually billions or more. This greatly magnifies any inefficiency in the calculations. Tensor processing units (TPUs) are purpose-built for AI/ML calculations and provide a valuable resource in the data center. However, they are too power intensive to support AI/ML in mobile applications, the edge, or IoT.

These applications require new approaches to power efficiency. *Machine Learning and Non-volatile Memories* explains the unique options for analog computing of matrix multiplications in emerging memory technologies. Chapters of this book detail how Phase Change Memory (PCM) and resistive RAM (RRAM) can be programmed to a range of resistance values, acting as weights in an analog implementation of Multiply-Accumulate (MAC) operations. It is widely believed that these emerging memory technologies will be especially important for practical implementations of neuromorphic computing.

AI/ML will also help improve yields, lower costs, and increase performance and reliability in a wide range of applications and systems. *Machine Learning and Non-volatile Memories* highlights AI/ML's flash management role in Solid-State Drives (SSDs). In particular, the 100+ layers of the latest 3D NAND flash chips can each have custom settings, optimized by AI/ML algorithms running on the controller of an SSD. This is especially important when using NAND chips that support four bits per cell (QLC—Quad Level Cell). Manufacturers are developing NAND capable of storing five bits per cell (PLC—Penta Level Cell) or more, for which such fine-tuning of parameters will be essential.

Machine Learning and Non-volatile Memories will join Rino Micheloni's previous books on storage and memory technologies as a trusted resource for professionals and academics. Anyone seeking clear, useful knowledge on the latest advances and next directions in storage for AI/ML will find them here. Thank you, Rino and all the contributing authors, for creating yet another valuable reference and guide for our industry.

January 2022

Charles H. Sobey
Chief Scientist, Channel Science, LLC
Plano, Texas, US

Acknowledgments

Writing a book requires a lot of time and energy; this is true for all books, but an engineering book includes so many graphs, numbers, and equations that make it prone to errors (or "bugs", as engineers would call them). This is the reason why we are so grateful to our colleagues who acted as reviewers: without them, such a complex book would have never reached its final form. Being engineers, we know that projects, despite all the efforts, might include bugs though…don't hesitate to reach out to us if you find any.

Machine Learning and, in more generic terms, Artificial Intelligence are very fascinating topics not only for engineers. Everybody can be affected by decisions taken by algorithms on behalf of human beings. As such, people from different fields can bring value to the table by bringing different views. A great example is the worldwide renowned sculptor, Alessandro Romano, who was so kind to give us permission to use a picture of his masterpiece "Medusa" to communicate the struggle of the human brain being replicated by a set of algorithms (this is how he envisioned Machine Learning in one of our conversations). We are so proud and thankful to have a picture of one of his masterpieces in our book.

Special thanks to Dr. Zach Evenson, Editor at Springer Nature, who helped us navigate through the challenges (i.e., delays) caused by the pandemic.

Last but not least, let us thank all the co-authors of the contributed chapters for their tireless dedication to this project: Thank You (we know you spent days and nights working on this project, on top of your normal duties).

Rino Micheloni
Cristian Zambelli

From the Editors

Machine Learning and Non-volatile Memories: Why Together?

At a first sight, Machine Learning and Non-volatile memories seem very far away from each other. Machine Learning implies mathematics, algorithms, and a lot of computation; non-volatile memories are solid-state devices used to store information, having the amazing capability of retaining the information even without power supply. This book will help the reader understand how these two worlds can work together, bringing a lot of value to each other. In particular, we can identify two main fields of application: analog Neural Networks (NNs) and Solid-State Drives (SSDs).

Let's start with the first field. Neural Networks are built to mimic the behavior of the human brain; to accomplish this result, NNs must perform a specific computation called Vector-by-Matrix (VbM) multiplication (Chapter "Neural Networks and Deep Learning Fundamentals"), which is particularly power hungry. In the digital domain, VbM is implemented by means of logic gates, which dictate both the area occupation and the power consumption; the combination of the two poses serious challenges to the hardware scalability, thus limiting the size of the neural network itself, especially in terms of the number of processable inputs and outputs. Non-volatile memories (PCM in chapter "Accelerating Deep Neural Networks with Phase-Change Memory Devices", RRAM in chapter "Analogue In-Memory Computing with Resistive Switching Memories", and Flash in chapter "Deep Neural Network Engines Based on Flash Technology") enable the analog implementation of the VbM (also called "neuromorphic architecture"), which can easily beat the equivalent digital implementation in terms of both speed and energy consumption. In 30 years of development [1], researchers have proven that the analog implementation, by adopting nanoscale solid-state devices, can even approach the power efficiency of the human brain [2–5].

The second field includes SSDs and the optimization of NAND flash memories in all their applications. Flash memories are extremely prone to errors; essentially, they are analog in nature because their ability of retaining the stored information is based on the capability of "locking" few electrons; just the temperature by itself is a good reason for electrons to leak away. Moreover, with the most recent 3D scaling (Chapter "Introduction to 3D NAND Flash Memories"), NAND Flash has become extremely difficult to handle as memory layers don't behave uniformly. As such, optimizing each single layer in terms of both electrical performances and reliability has become a real challenge as the number of parameters to be adjusted literally exploded. When looking at NAND components, the first usage model for Machine Learning is exactly the optimization of the available knobs, by learning from experiments on real silicon devices; for instance, Machine Learning can help grouping layers together, such that the same "recipe" can be applied to all layers within a group, thus reducing the overall design effort. The second model is related to SSDs. In fact, inside solid-state drives, NAND memories are managed by a microcontroller who has the task to execute all the necessary Flash management algorithms designed to extend the life of each NAND device. For a storage designer, it is difficult to directly access the internal architecture of the drive; therefore, going to a higher abstraction level (i.e., algorithms handling the data from/to the drive) is the only viable solution to improve the reliability and the performance of the drive; these algorithms are usually referred to as prognostics tools. Machine learning has emerged as a very promising candidate in the prognostics context because it offers advanced techniques to improve latency, throughput, reliability, and many other SSD's parameters with a minimal system overhead. Both usage models are covered in great detail in chapter "Machine Learning for 3D NAND Flash and Solid State Drives Reliability/Performance Optimization".

Let's now briefly take a closer look at the content of each chapter of this book.

Chapter "Introduction to Machine Learning" helps the reader understand the basics of machine learning, by analyzing some of the most popular learning models such as decision trees, random forest, and support vector machines. In a nutshell, chapter "Introduction to Machine Learning" introduces the concept of programming a computer to optimize one or more criteria, by using examples from past experiences.

When looking at machine learning tasks, Neural Networks, especially in the form of Deep Learning are the superstar. Chapter "Neural Networks and Deep Learning Fundamentals" provides a great introduction to the field, starting with a brief historical overview. Multilayer Perceptrons, Convolutional Neural Networks, and Recurrent Neural Networks are first explained and then used to show how they can mimic the human brain by looking at practical examples.

Given how ubiquitous deep Neural Networks are, the topic of accelerating their performances is of primary importance. In this context, analog computation, by means of analog memory devices, has shown an outstanding potential and, therefore, it is considered a research field of paramount importance. Chapter "Accelerating Deep Neural Networks with Phase-Change Memory Devices" describes how the multiply-accumulate operation (at the heart of NNs) can be performed in-memory, by using a Phase-Change Memory (PCM) as the resistive element within a crossbar array. Of course, nothing comes for free, and chapter "Accelerating Deep Neural Networks with Phase-Change Memory Devices" addresses the challenges associated with the usage of PCMs, including the impact of conductance drift on deep neural network accuracy.

PCM is not the only available option for analog computing. In chapter "Analogue In-memory Computing with Resistive Switching Memories", authors make use of a Resistive Switching Memory (RRAM), also known as memristor, to implement the in-memory computation. Controllable conductance, good scaling, and relatively low energy consumption are considered the key advantages of this technology. Chapter "Analogue In-Memory Computing with Resistive Switching Memories" provides a great overview of the RRAM-based implementation, going from the device level description to its electrical characteristics, and from the suitable computing architectures to the end applications.

Over the last few years, Solid-State-Drives (SSDs) have gained a lot of traction in the market, especially if equipped with very high-speed interfaces like PCIe/NVMe. At the beginning, SSDs spread into the consumer market, from laptops to tablets and smartphones. Now we can find SSDs in all sorts of enterprise applications, the SSD being one of the key storage elements in the most modern data centers.

In a nutshell, a Solid-State-Drive is made of a Flash microcontroller plus many NAND Flash memories; in high-speed applications, we are easily talking about hundreds of NAND dies soldered on a single SSD board. All these bits must be properly handled by the Flash microcontroller making the bridge between the System Host (i.e., a CPU with an Operating System) and the non-volatile memory subsystem.

Nowadays, as a matter of fact, NAND Flash memories are the key driver in the silicon process technology race. Today, we are talking about a 2Tb monolithic NAND die, i.e., 2 billion bits in less than 150 mm^2 of silicon. As always, this outstanding achievement is not coming for free. By nature, in fact, NAND Flash memories are very unreliable; considering how packed the memory cells are, this should not come as a surprise. To make NAND reliable enough, the Flash microcontroller runs a lot of signal processing algorithms (e.g., Error Correction Code, wear leveling, randomization, etc.), which require a lot of know-how during the design phase and consume time and power in the user's application.

Since 2016 the NAND complexity has grown even more, mainly because of the 3D architecture, which is the subject of chapter "Introduction to 3D NAND Flash Memories". Identifying the right way for going 3D was not so easy though and the chapter provides an overview of the main options for vertical scaling. 3D arrays can leverage either Floating Gate (FG) or Charge Trapping (CT) technologies and they are both reviewed in chapter "Introduction to 3D NAND Flash Memories".

With a good understanding of how Flash memories work, the reader will be ready to appreciate how they can be used to implement the Vector-by-Matrix (VbM) multiplication, which is the core of the hardware implementation of a Neural Network. Chapter "Deep Neural Network Engines Based on Flash Technology" deals with both NOR and NAND Flash memories, showing how they can be effectively used to build NNs in the analog domain. Indeed, the reader will understand how Flash memory cells, thanks to their tunable threshold voltage, can replicate the behavior of a synapse inside the human brain.

As already mentioned, SSDs and Flash memories are strictly coupled together and this is true in many applications, from consumer electronics to exa-scaled data centers; therefore, they must work together at their best capabilities. However, as 3D

Flash scales, there is a significant amount of work that must be done in order to optimize the overall performances of SSDs. Machine learning has emerged as a viable solution in many stages of this process. After introducing the main Flash reliability issues, chapter "Machine Learning for 3D NAND Flash and Solid State Drives Reliability/Performance Optimization" shows both supervised and unsupervised machine learning techniques used to identify homogeneous areas inside the Flash array, thus enabling an optimization of the storage system performance by means of a fine-tuned Error Correction Code. In addition, chapter "Machine Learning for 3D NAND Flash and Solid State Drives Reliability/Performance Optimization" deals with algorithms and techniques for a proactive reliability management of SSDs. Last but not least, chapter "Machine Learning for 3D NAND Flash and Solid State Drives Reliability/Performance Optimization" The last section of the chapter discusses the next challenge for machine learning in the context of SSDs for enterprise applications, namely the Computational Storage (CS) paradigm.

To conclude this introduction, let's get back to the questions we asked ourselves a few years ago.

Which problems Machine Learning could solve when looking at NAND Flash? There is a plethora of algorithms out there: which are the good ones for Flash? Machine learning or deep learning? How much can we gain in terms of useful memory lifetime by using Machine Learning? Can neuromorphic memories help? Can Machine Learning and Error Correction Codes work together in a more efficient way? And then the question with the capital Q: is it possible to develop an autonomous SSD, i.e., a drive that can optimize itself as working conditions in the field change?

We (editors and co-authors) tried our best to answer the above questions throughout the book. In some cases, answers are now clear, in some others, there is still a lot of work to be done. We do hope that this book can provide a good overview of the basic concepts and applications, helping engineers, data scientists, and researchers to bring the collaboration between Machine Learning and Non-volatile memories to the next level.

Enjoy!

Rino Micheloni
Cristian Zambelli

References

1. C. Mead, Analog VLSI and Neural Systems. Reading, MA, USA: Addison-Wesley, 1989.
2. G. Indiveri et al., Neuromorphic silicon neuron circuits. Frontiers Neurosci. **5**(73), 1–23, (2011)
3. K. Likharev, CrossNets: Neuromorphic hybrid CMOS/nanoelectronic networks. Sci. Adv. Mater. **3**, 322–331, (2011)
4. J. Hasler and H. Marr, Finding a roadmap to achieve large neuromorphic hardware systems. Frontiers Neurosci., **7**(118), (2013)
5. L. Ceze et al., Nanoelectronic neurocomputing: Status and prospects. In Proc. DRC, Newark, DE, USA, 1–2, (2016)

Introduction

Perhaps the most important revolution in the human history is the introduction of large-scale artificial intelligence. This revolution is having an incredible impact, beyond any imagination. Artificial intelligence will change how we design, conceive, make, and bring products to the market. Artificial intelligence and its applications will be able to change our life in many ways.

What is artificial intelligence, and why is it so revolutionary?

Let's start by saying that there are different types of artificial intelligence. The weak AI, the general AI, and the super AI called "superintelligence". All these types of artificial intelligence have in common the use of machines to think and act like human beings by identifying models and finding the description of the problem to be solved emulating cognitive processes similar to those used by humans in carrying out the same tasks.

As early as the 1600s, the English empiricist philosopher Thomas Hobbes said: "Reasoning is nothing but calculating" and Artificial Intelligence demonstrated that he was right. Even if the concept to use machine to emulate human brain capabilities is very old, we had to wait until 1956 before properly speaking of "Artificial Intelligence". The term was first coined and used by the American mathematician John McCarthy, during a seminar held in Dartmouth College, New Hampshire. The founders of this new discipline, John McCarty, Marvir Minsky, Allan Newell, Herbert Simon, and Claude Shannon (one of the fathers of cybernetics), believed that it was possible to use machines for voice recognition, understanding of natural language, identification of objects, deductive reasoning, discovering new mathematical theorems, and many other activities, previously a prerogative of the human mind only. The revolution started!

Today, the field of artificial intelligence is very broad and pervades various sectors: Games, Robotics, Images, Language Systems, Chatbots, Anti-Fraud Systems, Industrial Maintenance, Failure Prediction Systems, Cybersecurity Systems. Predictive medicine, advanced diagnostics, study of new drugs, advanced design, such as architecture, mechanics, electronics, and automotive.

But how does it work and why do we identify this type of machine behavior as intelligent? First of all, Artificial Intelligence is not a programming activity done by a programmer but a learning process performed by the machine itself. Artificial intelligence is based on a statistical comparison between "what it seems to be and what it is". It's a simple statistical comparison that can give rise to spectacular results that enable the reproduction of mechanisms related to the cognitive faculties of human beings by computer systems. A simple operation repeated many and many times makes computers learn the world around them!

If we think about how humans learn, in most cases, we discover that it is the statistical comparison that guides our knowledge, our intuition, our way of thinking. Indeed, there are entire disciplines of human knowledge that are based almost exclusively on this simple principle of statistical comparison. Let's think, for example, to medicine where causes and effects are associated trying to identify which are the components that most likely lead to a certain effect. Or we can talk about history, experimental physics, business, and military strategies. Everything, if we think about it, is based on a continuous comparison between causes and effects: we take the elements that we have identified as key points, we relate them to known effects, and from these analog and iterative processes, we formulate a theory that explains what we have observed.

On the same principle, it is possible to teach a machine to think and make decisions, with the simple operation of comparing reality and hypotheses, making the machine understand the hidden correlations and generate new hypotheses. When the hypothesis generated by the machine approaches or even coincides with what we know to be correct, it means that the machine has found a way to interpret what is happening correctly. To get more specific and understand the challenges to face when implementing artificial intelligence in the real world, we must understand the importance of computing power and computing architecture.

Computing power and architecture are two facts that must not be overlooked, comparable to the concept of having a gifted brain that works well, and having sufficient data or having access to knowledge to learn from. The above concepts work from the basic methods based on pure statistics, to the more advanced and complex ones based on neural networks and deep learning, also known as Machine Learning.

A neural network is, in substance, a concatenated series of identical operations based on simple functions, called artificial neurons, carried out on data. The neural networks take their name from the analogy with the human brain, where neurons are supposed to always perform the same type of operation. In both cases, the brain and the neural network, the connections between the various neurons determine the thought. The connections inside a neural network are the values assigned to the internal links. It is evident that, although each operation is relatively simple, having billions of linked operations is not a trivial situation, especially in terms of the required computing power.

To better understand what we are talking about, we can represent the simplest neural network model with three levels of processing: In the first, the input data are provided, in the second, data are processed by applying the aforementioned functions, and in the last level, the solutions are presented. Such a network does not have interesting intellectual abilities, it would be, so to speak, comparable to a brain with very few neurons, which, as we know, does not have great skills. By increasing the number of intermediate layers, called "hidden layers", you will get more neurons and obtain an enormously greater complexity which corresponds to the ability to be "more" intelligent and carry out much more complex thinking tasks. This kind of network requires a giant computational effort that was impracticable, due to lack of computing capacity, until a few years ago; other than that, this network requires having a vast and, above all, variegated database available to train the links of the network. In essence, the architecture and its related computing power are, therefore, crucial factors in obtaining satisfactory results.

The need for computing power opens up two paths for the future. First, the creation of new devices that allow to reach ever-higher computing powers, improve the data access speed, including new memory architectures, and deliver much more efficiency in data storing and loading. Second, the improvement of the computing performances will allow building increasingly sophisticated networks and algorithms. These next generations of algorithms will be capable of carrying out incredibly advanced tasks such as designing and managing complex systems creatively, using evolutionary genetic algorithms to define the best characteristics of a product, predicting complex future scenarios in many fields, from business to medicine, thus reaching unthinkable results with human intelligence alone.

In a very close future, we can imagine machines that design machines. Artificial Intelligence systems, soon, will be able to invent and design components used to build even better and more efficient machines. We can imagine new algorithms making the machines work more efficiently, and helping us to save money and resources.

This book opens the way to understanding the principles on which we build machine learning-based applications. It makes us understand the importance of decisive aspects in the architecture of this revolutionary discipline. The book helps us understand "Machine Learning" and its related challenges, scenarios, and opportunities, especially in the context of non-volatile memories. It's a must-read book for those who want to be a protagonist of the future and not a simple spectator.

Emilio Billi
A3Cube CTO and AI Guru

Contents

Editors and Contributors

About the Editors

Dr. Rino Micheloni is a Research Fellow at the University of Ferrara, Italy. Before that, he was Vice-President and Fellow at Microsemi/Microchip Corporation, where he established the Flash Signal Processing Labs in Milan, Italy, with special focus on NAND Flash technology characterization, Machine Learning techniques for improving memory reliability, and Error Correction Codes. Prior to joining Microsemi, he was a Fellow at PMC-Sierra, working on NAND Flash technology characterization, LDPC, and NAND Signal Processing as part of the team developing Flash controllers for PCIe SSDs. Before that, he was with IDT (Integrated Device Technology) as Lead Flash Technologist, driving the architecture and design of the BCH engine in the world's first PCIe NVMe SSD controller. Early in his career, he led NAND design teams at STMicroelectronics, Hynix, and Infineon; during this time, he developed the industry's first MLC NOR device with embedded ECC technology and the industry's first MLC NAND with embedded BCH.

Dr. Micheloni is IEEE Senior Member, he has co-authored 100 publications in peer-reviewed journals and international conferences, and he holds 291 patents worldwide (including 139 US patents). He received the STMicroelectronics Exceptional Patent Award in 2003 and 2004, the Infineon IP Award in 2007, and he was elected to the PMC-Sierra Inventor Wall of Fame in 2013. In 2020, Dr. Micheloni was selected for the European Inventor Award.

Dr. Micheloni has published the following books: 3D Flash Memories (Chinese edition by Tsinghua University Press, 2020), Inside Solid State Drives—2nd edition—(Springer, 2018), Solid-State-Drives (SSDs) Modeling (Springer, 2017), 3D Flash Memories (Springer, 2016), Inside Solid State Drives (Springer, 2013), Inside NAND Flash Memories (Springer, 2010), Error Correction Codes for Non-volatile Memories (Springer, 2008), Memories in Wireless Systems (Springer, 2008), VLSI-Design of Non-volatile Memories (Springer, 2005), and Memorie in Sistemi Wireless (Franco Angeli, 2005).

Dr. Cristian Zambelli received the M.Sc. and the Ph.D. degrees in Electronic Engineering from the University of Ferrara, Ferrara, Italy, in 2008 and 2012, respectively. Since 2015, he holds an Assistant Professor position with the same institution where he teaches Data Storage, Electronic Systems Reliability, and FPGA Laboratory. His current research interests include the electrical characterization, physics, and reliability modeling of different non-volatile memories such as NAND/NOR Flash planar and 3D integrated, Phase Change Memories, Nano-MEMS memories, Resistive RAM (RRAM), and Magnetic RAM. He is also interested in the evaluation of the Solid State Drives reliability/performance trade-offs exposed by the integrated memory technology and the implementation of Computational Storage paradigms by exploiting FPGAs and GPUs.

Dr. Zambelli is an IEEE Member and authored or co-authored up to 107 publications in peer-reviewed journals, proceedings of international conferences, and book chapters. He serves in the TPC of several international conferences like IRPS, ESSDERC/ESSCIRC, and IIRW being for the latter also in the management committee. He is currently serving as Associate Editor for the IEEE Access journal since 2019.

Contributors

Stefano Ambrogio IBM Research-Almaden, San Jose, CA, USA

Elena Bellodi Department of Engineering, University of Ferrara, Ferrara, Italy

Geoffrey W. Burr IBM Research-Almaden, San Jose, CA, USA

An Chen IBM Research-Almaden, San Jose, CA, USA

Luca Crippa Busnago (MB), Italy

Andrea Fasoli IBM Research-Almaden, San Jose, CA, USA

Michele Fraccaroli Department of Engineering, University of Ferrara, Ferrara, Italy

Alexander Friz IBM Research-Almaden, San Jose, CA, USA

Daniele Ielmini Dipartimento di Elettronica, Informazione e Bioingegneria, and IU.NET, Politecnico di Milano, Milano, Italy

Evelina Lamma Department of Engineering, University of Ferrara, Ferrara, Italy

Charles Mackin IBM Research-Almaden, San Jose, CA, USA

Rino Micheloni Dipartimento di Ingegneria, Università degli Studi di Ferrara, Ferrara, Italy

Pritish Narayanan IBM Research-Almaden, San Jose, CA, USA

P. Olivo Dipartimento di Ingegneria, Università degli Studi di Ferrara, Ferrara, Italy

Giacomo Pedretti Dipartimento di Elettronica, Informazione e Bioingegneria, and IU.NET, Politecnico di Milano, Milano, Italy

Fabrizio Riguzzi Department of Mathematics and Computer Science, University of Ferrara, Ferrara, Italy

Katie Spoon IBM Research-Almaden, San Jose, CA, USA

Hsinyu Tsai IBM Research-Almaden, San Jose, CA, USA

Cristian Zambelli Dipartimento di Ingegneria, Università degli Studi di Ferrara, Ferrara, Italy

Riccardo Zese Department of Engineering, University of Ferrara, Ferrara, Italy; Department of Chemical, Pharmaceutical and Agricultural Sciences, University of Ferrara, Ferrara, Italy

Introduction to Machine Learning

Elena Bellodi, Riccardo Zese, Fabrizio Riguzzi, and Evelina Lamma

Abstract Machine learning is programming computers to optimize a performance criterion using example data or past experience. We need learning in cases where we cannot directly write a computer program to solve a given problem, but need example data or experience. Another case is when the problem to be solved changes in time, or depends on the particular environment. In this Chapter we introduce the basic components of machine learning and focus on a few very popular machine learning models: decision trees, random forest (tree models) and support vector machines (geometric and linear models).

1 Overview

Machine learning (ML) is the systematic study of algorithms and systems that improve their *knowledge* or *performance* with experience. Experience may take different forms, such as labelled training data, corrections of mistakes, rewards when a certain goal is reached, among many others. Machine learning algorithms may be directed at improving performance on a certain task, typically their ability to recognize *future* data, but may more generally result in improved knowledge.The main

E. Bellodi (✉) · R. Zese · E. Lamma
Department of Engineering, University of Ferrara, Via Saragat 1, 44122 Ferrara, Italy
e-mail: elena.bellodi@unife.it

E. Lamma
e-mail: evelina.lamma@unife.it

R. Zese
Department of Chemical, Pharmaceutical and Agricultural Sciences, University of Ferrara, Via Luigi Borsari, n. 46, 44121 Ferrara, Italy
e-mail: riccardo.zese@unife.it

F. Riguzzi
Department of Mathematics and Computer Science, University of Ferrara, Via Saragat 1, 44122 Ferrara, Italy
e-mail: fabrizio.riguzzi@unife.it

R. Micheloni and C. Zambelli (eds.), *Machine Learning and Non-volatile Memories*,
https://doi.org/10.1007/978-3-031-03841-9_1

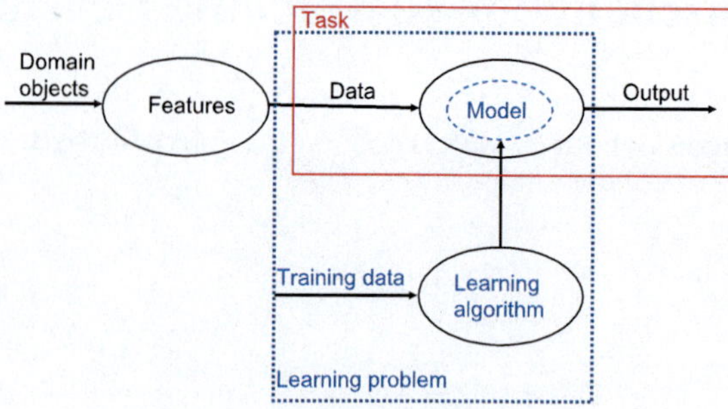

Fig. 1 An overview of how machine learning is used to address a given task [1]. A task (solid-line box) requires an appropriate mapping—a model—from data described by features to outputs. Obtaining such a mapping from training data is what constitutes a learning problem (dashed-line box)

ingredients of machine learning are: tasks, models and features [1]. What is called a ML application is the construction of a model that solves a practical task, by means of machine learning methods, using data from the task domain.

Suppose we have a large 'training set' of e-mails which have been hand-labelled spam or ham, and we know the results of all the tests for each of these e-mails. The goal is now to come up with a weight for every test, such that all spam e-mails receive a score above 5, and all ham e-mails get less than 5. This is an example of *task*. Given this task, we want an algorithm that learns to recognize spam e-mail from examples and counter-examples. Moreover, the more 'training' data is made available, the better the algorithm will become at this task:

- We call this type of task *binary classification*, as it involves assigning objects (e-mails) to one of two classes: spam or ham;
- This task is achieved by describing each e-mail in terms of a number of variables or *features*;
- We have to figure out a connection between the features and the class—we call such a connection a *model*—by analyzing a training set of e-mails already labelled with the correct class.

Fig. 1 shows how these ingredients relate. The model is produced as the output of a machine learning algorithm applied to training data; there is a wide variety of models to choose from, and we will investigate some of them in this chapter. No matter what variety of machine learning models one may consider, they are designed to solve only a small number of tasks and use only a few different types of features.

In the spam filtering example, the model could be a linear equation of the form $\sum_{i=1}^{M} w_i x_i > t$, where the x_i denote the 0–1 valued or 'Boolean' features indicating whether the i-th test succeeded for the e-mail, w_i are the feature weights learned from the training set, and t is the threshold above which e-mails are classified as spam.

This chapter dedicates a section for each of the three ML ingredients. Section 2 describes what machine learning means by features'. Section 3 presents the types of problems that can be solved with machine learning (tasks). Sections 4 and 5 focus on two kinds of machine learning models, namely tree models (Decision Trees and Random Forest) and linear models (in particular Support Vector Machines).

2 Features

In machine learning, data are often referred to as 'examples' or 'instances'. *Features*, also called attributes, are defined as mappings $f_i : \mathcal{X} \to \mathcal{F}_i$ from the instance space $\mathcal{X} = \mathbb{R}^d$ to the feature domain $\mathcal{F}_i$, where $\mathbb{R}^d$ is the d-dimensional Euclidean space [1].

We can distinguish features by their domain: common feature domains include real and integer numbers, but also discrete sets such as colours, Booleans, and so on. We can also distinguish features by the range of permissible operations. Although many data sets come with pre-defined features, they can be manipulated in many ways, a process that is called *feature transformation*. For example, we can change the domain of a feature by rescaling or discretization; we can select the best features from a larger set and only work with the selected ones; or we can combine two or more features into a new feature. Sometimes data do not come with built-in features (for instance in text classification), so they need to be constructed by the developer of the machine learning application: this *feature construction* process is crucial for the success of the application.

3 Learning Tasks

Let us give symbols and names to the main components of ML [2]:

- The instance space $\mathcal{X}$, also called input space, is the space of all possible inputs $\mathbf{x}$;
- $\mathcal{Y}$ is the output space, the set of all possible outputs of the model;
- There exists an unknown target function $g : \mathcal{X} \to \mathcal{Y}$, the ideal formula relating each input to each output;
- There is a data set $\mathcal{D}$ of N input examples $(\mathbf{x}_1, ..., \mathbf{x}_N)$: inputs are the training data available to the machine learning algorithm;
- The learning algorithm uses $\mathcal{D}$ to pick a formula $\hat{g} : \mathcal{X} \to \mathcal{Y}$ that approximates g. The algorithm chooses $\hat{g}$ from a set of candidate formulas under consideration, which we call the hypothesis set $\mathcal{H}$;
- The formula $\hat{g}$ represents the model that the learning algorithm produces; $\hat{g}$ is chosen in order to best match g on the *training* examples available to the algorithm, with the hope that it will continue to match g on new ones, called 'test' or 'unseen' examples.

When the training data contains explicit examples of what the correct output should be for given inputs, i.e., it contains couples $(\mathbf{x}_1, y_1), ..., (\mathbf{x}_N, y_N)$ where $y_n = g(\mathbf{x}_n)$, then we are within the *supervised learning task*: y_n is called the 'target variable' or 'label'. *Classification* and *regression* assume the availability of a training set of examples labelled with true classes or function values respectively, and will be investigated in Sect. 3.1.

Providing the true labels for a data set is often labour-intensive and expensive. *Semi-supervised learning* is the branch of machine learning concerned with using labelled as well as unlabelled data: it permits taking advantage of the large amounts of unlabelled data available in many use cases in combination with typically smaller sets of labelled data [3]. A large majority of the research on semi-supervised learning is focused on classification.

In contrast to supervised learning where the training examples are of the form (input, correct output), in *reinforcement learning* they are of the form (input, some output, grade for this output), where the grade is a measure of how good that output is. Reinforcement learning is typically the training of machine learning models to make a sequence of decisions subject to rewards or penalties for the actions an agent performs in the environment.

In an *unsupervised learning task*, we are just given input examples $\{\mathbf{x}_1, ..., \mathbf{x}_N\}$, which represent 'unlabelled data'. Unsupervised learning can be viewed as the task of spontaneously finding patterns and structure in input data:

- The task of grouping data without prior information on the groups is called *clustering*; a clustering algorithm works by assessing the similarity between instances (e.g., documents belonging to similar subjects) and putting similar instances in the same cluster (representing the same subject) and 'dissimilar' instances in different clusters;
- *Association rules* are a kind of pattern that are popular in marketing applications, and are found by data mining algorithms that search for items that frequently occur together.

A parallel distinction to that between supervised and unsupervised learning is the distinction between whether the model output involves the target variable or not: we call it a *predictive model* if it does, and a *descriptive model* if it does not. This leads to the four main different machine learning settings [1]:

1. Supervised learning of predictive models;
2. Supervised learning of a descriptive model (called subgroup discovery: the model identifies subsets of the data that behave differently with respect to the target variable);
3. Unsupervised learning of a descriptive model (clustering, association rules, matrix decomposition);
4. Unsupervised learning of a predictive model (we cluster data with the intention of using the clusters to assign class labels to new data).

3.1 Supervised Learning of Predictive Models

Classification and regression tasks fall into the category of supervised learning of predictive models.

A **classifier** is a mapping $g : \mathcal{X} \rightarrow \mathcal{C}$, where $\mathcal{C} = \{C_1, C_1, ..., C_k\}$ is a finite and usually small set of class labels. Examples for a classifier take the form $(\mathbf{x}, y)$, where y is the *true class* of the instance. In the simplest case we have only two classes which are usually referred to as positive and negative, 1 and 0, or +1 and -1. Two-class classification is often called *binary classification*. Spam e-mail filtering is a good example of binary classification, in which spam is conventionally taken as the positive class, and ham as the negative class. If we have more than two classes we are dealing with *multi-class classification*.

The task of the classifier is to assign an unseen input $\mathbf{x}$ to one of the classes: this assignment represents the prediction. There is a considerable range of machine learning classifiers to choose from: linear classifiers, Support Vector Machines (SVM), Naive Bayes, Decision trees (DT) and Random forests (RF), k-Nearest Neighbors (KNN), Neural Networks. DT and RF will be discussed in Sect. 4, SVM in Sect. 5.

The output space $\mathcal{Y}$ may not be a discrete set of classes. A **regressor** is a mapping $g : \mathcal{X} \rightarrow \mathbb{R}$, i.e. the target variable y is real-valued. Let us say we want to have a system that can predict the price of an used car. Inputs are the car attributes - brand, year, engine capacity, mileage, and other information - that we believe affect a car's worth. The output is the price of the car. The machine learning algorithm, again surveying the past transactions $\mathbf{x}$, will fit an 'estimator function' to this data to learn y (the price) as a function of the car attributes. *Linear regression* is a linear model, meaning that g is a linear combination of input features, with weights applied to each feature:

$$y = g(x \mid w_1, w_0) = w_1 x + w_0$$

in the case of a single feature (univariate regression), or more generally:

$$y = g(x \mid w_k, ..., w_2, w_1, w_0) = w_k x^k + ... + w_2 x^2 + w_1 x + w_0 = \mathbf{w}^\mathbf{T}\mathbf{x} = \mathbf{w} \cdot \mathbf{x}$$

The problem is to come up with good values for the weights - ones that make the model's output match the desired output. Here, the output and the inputs - attribute values - are all numeric. Linear models are the easiest to visualize in two dimensions, where they correspond to drawing a straight line through a set of data points, which represents the prediction equation.

Note that a linear regressor can be reduced to a linear classifier if y is thresholded at zero to produce a ± 1 output, appropriate for binary decisions, i.e. $y = sign(\mathbf{w}^\mathbf{T}\mathbf{x})$ and the line in two dimensions graphically divides the examples of the two classes.

The least-squares method can be used to learn the weights of a linear regressor by considering the differences between the actual and estimated function values on the training examples, called residuals $\epsilon_i = g(x_i) - \hat{g}(x_i)$. The method consists in finding $\hat{g}$ such that $\sum_{i=1}^{n} \epsilon_i^2$ over all examples is minimized.

Logistic regression is a linear model that outputs a probability, a value between 0 and 1. It has similarities to both previous models, as the output is real (like regression) but bounded (like classification). Suppose we want to predict the occurrence of heart attacks based on a person's cholesterol level, blood pressure, age, weight, and other factors. Obviously, we cannot predict a heart attack with any certainty, but we may be able to predict how likely it is to occur given these factors. The closer y is to 1, the more likely that the person will have a heart attack. The target variable is defined as

$$y = g(x) = \sigma(\mathbf{w}^\mathbf{T}\mathbf{x})$$

where σ is the so-called logistic function $\sigma(s) = \frac{e^s}{1+e^s}$ whose output is between 0 and 1. It is also called a sigmoid because its shape looks like a flattened out ‘s’. The output can be interpreted as a probability for a binary event (heart attack or no heart attack, etc.).

4 Tree Models

Tree models are among the most popular models in machine learning. Trees are expressive and easy to understand by humans, are more explainable than mathematical models and of particular appeal to computer scientists due to their recursive ‘divide-and-conquer’ nature. Tree models can be employed to solve many machine learning tasks, including classification, regression and clustering. The tree structure that is common to all those models can be defined as follows [1].

Definition 1 (Literal) Literals are logical expressions representing equalities of the form *Feature* = *Value* and, for numerical features, inequalities of the form *Feature* > *Value*.

Definition 2 (Feature tree) A feature tree is a tree such that each internal node (the nodes that are not leaves) is labelled with a feature, and each edge emanating from an internal node is labelled with a literal. The set of literals at a node is called a *split*. Each leaf of the tree represents a logical expression, which is the conjunction of literals encountered on the path from the root of the tree to the leaf.

Algorithm 1 [1] gives the generic learning procedure common to most tree learners, called ‘tree induction’. It is a divide-and-conquer algorithm: it divides the data into subsets, builds a tree for each of those and then combines those subtrees into a single tree.

$Homogeneous(D)$ returns true if the instances in data D are homogeneous enough to be labelled with a single label, and false otherwise; $Label(D)$ returns the most appropriate label for a set of instances D; $BestSplit(D, F)$ returns the best set of literals to be put at the root of the tree.

Note that the tree structure is not fixed a priori but the tree grows, branches and leaves are added during learning depending on the complexity of the problem inherent in the data. For a given training set, there exist many trees that code it with no error, and, for simplicity, we are interested in finding the smallest among them, where tree size is measured as the number of nodes in the tree and the complexity of the decision nodes. Finding the smallest tree is NP-complete[1], and we are forced to use

[1] In computational complexity theory, **NP** is the class of problems that can solved in **P**olynomial time by a **N**ondeterministic Turing machine. NP-complete problems are the maximally difficult problems in NP and no efficient solution algorithm has been found for them. For efficient algorithms we mean

Algorithm 1 $GrowTree(D, F)$ - grow a feature tree from training data.

```
Require: data D; set of features F
Ensure: feature tree T with labelled leaves
if Homogeneous(D) then
    return Label(D);
end if
S ← BestSplit(D, F);
split D into subsets D_i according to the literals in S;
for each i do
    if D_i ≠ ∅ then
        T_i ← GrowTree(D_i, F)
    else
        T_i is a leaf labelled with Label(D);
    end if
end for
return a tree whose root is labelled with S and whose children are T_i
```

local search procedures based on heuristics that give reasonable trees in reasonable time. Such algorithms are greedy: whenever there is a choice (such as choosing the best split), the best alternative is selected on the basis of the information then available, and this choice is never reconsidered [4].

In the remainder of this section we will instantiate the generic Algorithm 1 to classification, regression and clustering tasks.

4.1 Decision Trees

A decision tree is composed of internal *decision nodes* and terminal leaves.

For a **classification task** we can simply define a set of instances D to be homogenous if they are all from the same class, and the function $Label(D)$ will then obviously return that class. Now we have to define function $BestSplit(D, F)$.

In a *univariate* tree, $BestSplit(D, F)$ uses only one of the input features F in each internal node. If the used input feature x_j is discrete, taking one of n possible values, the node checks the value of x_j and takes the corresponding branch, implementing an n-way split. For example, if an attribute is $color \in \{red, blue, green\}$, then a node on that attribute has three children, each one corresponding to one of the three possible literals $color = red, color = blue, color = green$. As an internal node has discrete branches, a numeric input should be discretized. If x_j is numeric (ordered), $BestSplit(D, F)$ makes a comparison corresponding to the literal $x_j > w_m$ for node m, where w_m is a suitably chosen threshold value. The node m divides the input space into two: $L_m = \{\mathbf{x} \mid x_j \leq w_m\}$ and $R_m = \{\mathbf{x} \mid x_j > w_m\}$.

algorithms that run in polynomial time, because exponential-time algorithms have execution times that grow too rapidly as the problem size increases.

The goodness of a split is quantified by an 'impurity measure'. A split is 'pure' if after the split, for all branches, all the instances choosing a branch belong to the same class. If the split is pure, we do not need to split any further and the children are leaf nodes labelled with the class C_i of its examples. Possible functions to measure impurity are entropy [4], Gini index [5] or misclassification error. Research has shown that there is not a significant difference between these three measures. This is the basis of the 'Classification and Regression Trees' (CART) algorithm [5], ID3 algorithm [4], and its extension C4.5 [6]. When there is noise, i.e. mislabelled instances, growing the tree until it is purest could lead to a very large tree and it may overfit (it fits too much the specific training data from which it is built): so tree construction ends when nodes become pure enough, namely, a subset of data is not split further if impurity is lesser than a threshold.

Figure 2a shows a dataset of labor negotiations for 40 contracts: each instance concerns one contract, and the outcome is whether the contract is deemed acceptable ($class = good$) or unacceptable ($class = bad$). Figure 2b shows a learnt decision tree for classification of new contracts.

For a **regression task** each leaf node is labelled with a numeric value instead of a class. A leaf node defines a localized region in the input space where instances falling in this region have very similar numeric outputs. A regression tree is constructed in almost the same manner as a classification tree, except that the impurity measure, which is appropriate for classification, is replaced by a measure appropriate for regression: the goodness of a split is measured by the mean square error from the estimated value g_m in node m, calculated as the mean of the outputs of instances reaching the node. If at a node the error is acceptable, then a leaf node is created and it stores the g_m value. An example of regression tree is shown in Fig. 3.

The simple kind of regression tree considered here also suggests a way to learn **clustering** trees [1], even if regression is a supervised learning problem while clustering is unsupervised. We can introduce an abstract function $Dis : \mathcal{X} \times \mathcal{X} \rightarrow \mathbb{R}$ that measures the distance or dissimilarity of any two instances $\mathbf{x}, \mathbf{x}' \in \mathcal{X}$, such that the higher $Dis(\mathbf{x}, \mathbf{x}')$ is, the less similar $\mathbf{x}$ and $\mathbf{x}'$ are. The cluster dissimilarity of a set of instances D is then calculated as

$$Dis(D) = \frac{1}{|D|^2} \sum_{\mathbf{x} \in D} \sum_{\mathbf{x}' \in D} Dis(\mathbf{x}, \mathbf{x}')$$

The weighted average cluster dissimilarity over all children of a split gives the split dissimilarity, which can be used to inform $BestSplit(D, F)$. The label of a cluster may be its most representative instance, for example the one whose total dissimilarity to all other instances is lowest.

Pruning Frequently, a node is not split further if the number of training instances reaching a node is smaller than a certain percentage of the training set - for example, 5% - regardless of the impurity or error, as any decision based on too few instances causes variance. Stopping tree construction early on is called *prepruning* the tree. Another possibility is *postpruning*: we grow the tree until all leaves are pure, we then

Attribute	Type	1	2	3	...	40
Duration	(number of years)	1	2	3		2
Wage increase 1st year	Percentage	2%	4%	4.3%		4.5
Wage increase 2nd year	Percentage	?	5%	4.4%		4.0
Working hours per week	(number of hours)	28	35	38		40
Pension	{none, ret-allw, empl-cntr}	None	?	?		?
Statutory holidays	(number of days)	11	15	12		12
Health plan contribution	{none, half, full}	None	?	Full		Half
Acceptability of contract	{good, bad}	Bad	Good	Good		Good

(a) Labor negotiations dataset. Column 'Attribute' lists the input features, column 'Type' the feature type, columns 1-40 specify the values taken by 40 instances for each feature. Last row in the table specifies the class.

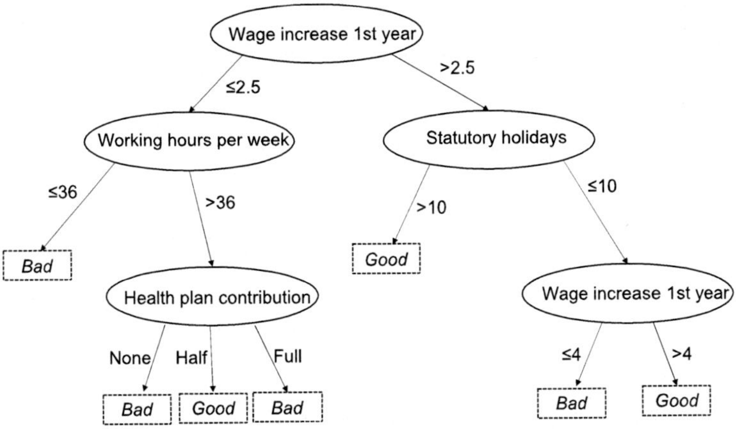

(b) A decision tree for classification learnt from the dataset.

Fig. 2 A dataset (top) with 40 instances and 2 classes (good, bad) and a decision tree (bottom) learnt from it

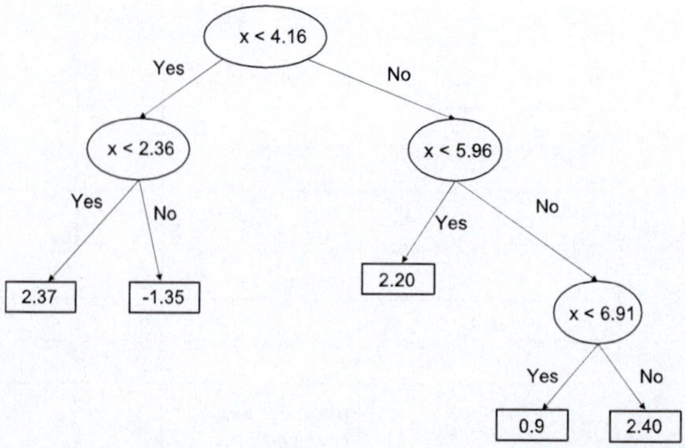

Fig. 3 A regression tree

find subtrees that cause overfitting and prune them. From the initial labelled set, we set aside a pruning set, unused during training. For each subtree, we replace it with a leaf node labelled with the majority class of the training instances covered by the subtree. If the leaf node does not perform worse than the subtree on the pruning set, we prune the subtree and keep the leaf node because the additional complexity of the subtree is not justified; otherwise, we keep the subtree. Prepruning is faster but postpruning generally leads to more accurate trees.

Rule Extraction Another main advantage of decision trees is *interpretability*: decision nodes carry conditions that are simple to understand. Each path from the root to a leaf corresponds to one conjunction of literals that can be written down as a set of *IF-THEN rules*, called a 'rule base'. One such method is "C4.5Rules" [6]. For example, the decision tree of Fig. 2b can be written down as the following set of rules:

```
R1: IF (Wage increase 1st year ≤ 2.5%) AND (Working hours per week > 36) THEN Contract=Bad
R2: IF (Wage increase 1st year ≤ 2.5%) AND (Working hours per week > 36)
    AND (Health plan contribution=None) THEN Contract=Bad
R3: IF (Wage increase 1st year ≤ 2.5%) AND (Working hours per week > 36)
    AND (Health plan contribution=Half) THEN Contract=Good
R4: IF (Wage increase 1st year ≤ 2.5%) AND (Working hours per week > 36)
    AND (Health plan contribution=Full) THEN Contract=Bad
R5: IF (Wage increase 1st year > 2.5%) AND (Statutory holidays > 10) THEN Contract=Good
R6: IF (Wage increase 1st year > 2.5%) AND (Statutory holidays > 10)
    AND (Wage increase 1st year ≤ 4) THEN Contract=Bad
R7: IF (Wage increase 1st year > 2.5%) AND (Statutory holidays > 10)
    AND (Wage increase 1st year > 4) THEN Contract=Good
```

Feature Use A univariate tree only uses the necessary variables, and after the tree is built, certain features may not be used at all. We can also say that features closer to the root are more important globally. It is possible to use a decision tree for *feature extraction*: we build a tree and then take only those features used by the tree as inputs to another learning method.

In a *multivariate* tree, at a decision node, all input dimensions can be used and thus it is more general. When all inputs are numeric, a binary linear multivariate node m is defined as $\mathbf{w}_\mathbf{m}^\mathbf{T}\mathbf{x} + w_{m0} > 0$, i.e. a weighted linear sum. Discrete attributes should be represented by 0/1 dummy numeric variables. It is possible to make it even

more flexible by using a nonlinear multivariate node. For example, for a quadratic node, we have $\mathbf{x^T W_m x} + \mathbf{w_m^T x} + w_{m0} > 0$.
The earliest algorithm proposed for learning multivariate decision trees for classification is the multivariate version of the CART algorithm [5], which fine-tunes the weights w_{mj}. CART also has a preprocessing stage to decrease dimensionality through subset selection and to reduce the complexity of the node. An evolution of CART is the OC1 algorithm [7]. Linear multivariate nodes are more difficult to interpret.

The *omnivariate* decision tree [8] is a hybrid tree architecture where the tree may have univariate, linear multivariate, or nonlinear multivariate nodes. Results show that more complex nodes are used early in the tree, closer to the root, and as we go down the tree, simple univariate nodes suffice. As we get closer to the leaves, we have simpler problems and, at the same time, we have less data.

Decision trees are used more frequently for classification than for regression. It is even the case that a decision tree is preferred over more accurate methods, because it is interpretable. When written down as a set of IF-THEN rules, the tree can be understood and the rules can be validated by human experts who have knowledge of the application domain.

Another big advantage of the univariate tree is that it can use numeric and discrete features together, without needing to convert one type into the other.

A decision tree, once it is constructed, does not store all the training set but only the structure of the tree, the parameters of the decision nodes, and the output values in leaves; this implies that the space complexity is very low.

4.2 Random Forests

Combinations of models are generally known as *model ensembles*. They are among the most powerful techniques in machine learning, often outperforming other methods. Random forests are an ensemble learning method for classification that grows many classification trees.

A random forest is a classifier consisting of a collection of tree-structured classifiers $g_k(\mathbf{x})$, $k = 1, \ldots$. where each tree casts a unit vote for the most popular class at input $\mathbf{x}$ [9]. Each tree is grown as follows (see Fig. 4a):

1. If the number of examples in the training set is N, sample N examples at random but with replacement, from the original data (procedure called ‘bagging’). This sample is known as ‘bootstrap sample’ and will be the training set for growing the tree. The bootstrap sample will in general contain duplicates, and hence some of the original data points will be missing even if the bootstrap sample is of the same size as the original data set. This is exactly what we want, as differences between the bootstrap samples will create diversity among the models in the ensemble. When the training set for the current tree is drawn, about 1/3 of the examples are

left out of the sample. This out-of-bag (oob) data is used to get a running unbiased estimate of the classification error as trees are added to the forest.
The use of bagging seems to enhance accuracy of the final model; also, bagging can be used to give ongoing estimates of the generalization error (the error on new examples) of the ensemble of trees;
2. If there are F input features, a number $f \ll F$ is specified such that at each node, f features are selected at random out of the F ('random feature selection') and the best split on these f is used to split the node. The value of f is maintained fixed during the forest growing. This is the only adjustable parameter to which random forests are sensitive;
3. Each tree is grown to the largest extent possible and there is no pruning. Random forests do not overfit as more trees are added.

In random forests, there is no need for cross-validation or a separate test set to get an unbiased estimate of the test set error. It is estimated internally, during the run, as follows:

- Each tree is constructed using a different bootstrap sample from the original data. About one-third of the examples are left out of the bootstrap sample and not used in the construction of the k-th tree.
- Each example left out in the construction of the k-th tree is passed down through the k-th tree to get a classification ('vote'). In this way, a test set classification is obtained for each example in about one-third of the trees.
- At the end of the run, these votes are aggregated to obtain the majority class j: the proportion of times that j is not equal to the true class of each example averaged over all examples is the oob error estimate.

To classify a new input vector $\mathbf{x}$, it is enough to pass the vector down each of the trees in the forest. Each tree gives a classification, and we say the tree 'votes' for that class. The forest chooses the classification having the most votes over all the trees in the forest (Fig. 4b).

In unsupervised learning the data consist of a set of $\mathbf{x}$ vectors of the same dimension with no class labels. The approach in random forests is to consider the original data as class 1 and to create a synthetic second class of the same size that will be labelled as class 2, created by randomly sampling from the univariate distributions of the original data. Here is how a single member of class two is created: the first coordinate is sampled from the N values taken by the first coordinate of the N examples in the data set; the second coordinate is sampled independently from the N values taken by the second coordinate of the N examples, and so forth. This artificial two-class problem can be run through random forests. This allows all of the random forests options to be applied to the original unlabelled data set.

In some data sets, the prediction error between classes is highly unbalanced. Some classes have a low prediction error, others a high one. This occurs usually when one class is much larger than another. Random forests will keep the error rate low on the large class while letting the smaller classes have a larger error rate to try to minimize the overall error rate.

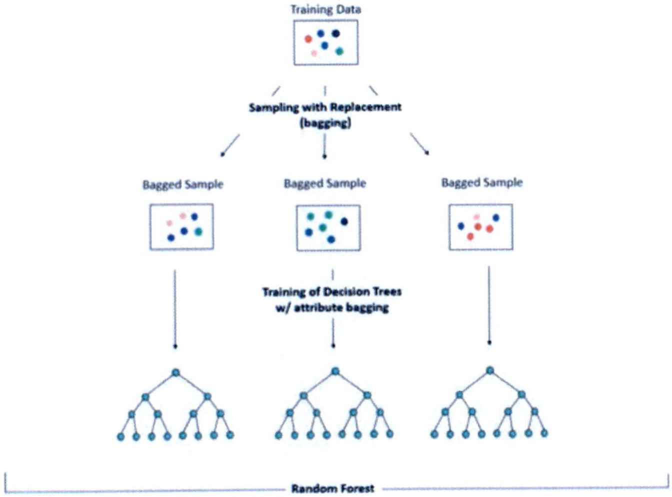

(a) Training phase.

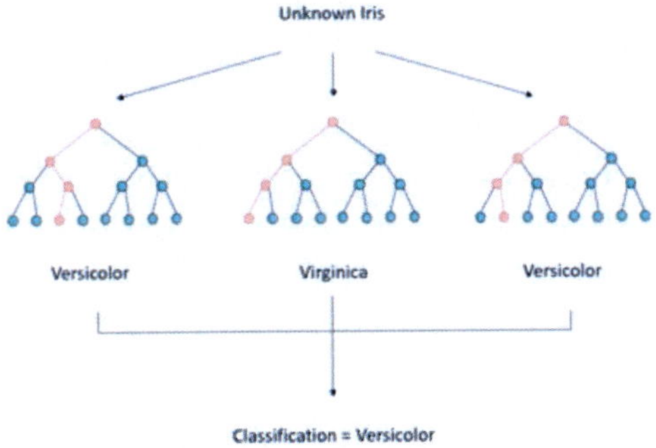

(b) Classification phase.

Fig. 4 Representation of training and classification for RF

Other advantages in using this method are:

- It runs efficiently on large databases;
- It can handle thousands of input features;
- It gives estimates of what features are important in the classification;
- It has an effective method for estimating missing data and maintains accuracy when a large proportion of the data are missing.

5 Linear Models

Linear models are a different approach for classification, probability estimation and regression: they are defined in terms of the geometry of instance space. We can then use geometric concepts such as lines and planes to impose structure on this space.

Linear models are 'parametric', meaning that they have a fixed form with a small number of numeric parameters that need to be learned from data. This is different from tree models, where the structure of the model (e.g., which features to use in the tree, and where) is not fixed in advance. Linear models are stable, meaning that small variations in the training data have only limited impact on the learned model. Tree models tend to vary more with the training data, as the choice of a different split at the root of the tree typically means that the rest of the tree is different as well. Linear models are less likely to overfit the training data than some other models, largely because they have relatively few parameters. The flip side of this is that they sometimes lead to *underfitting*, i.e., the machine learning model is not complex enough to accurately capture relationships between dataset's features and a target variable.

If all features are numerical, then we can use each feature as a coordinate in a Cartesian coordinate system, as they are easy to visualize, as long as we keep to two or three dimensions. It is important to keep in mind, though, that a Cartesian instance space has as many coordinates as there are features, which can be tens, hundreds, thousands, or even more. Such high-dimensional spaces are nevertheless very common in machine learning. Geometric concepts that potentially apply to high-dimensional spaces are usually prefixed with 'hyper-'.

We dedicate this section to looking in detail at the **support vector machine (SVM)**, also called 'maximum margin classifier', which became popular some years ago for solving problems in classification, regression, and novelty detection.

Let's recall the classifier for spam e-mail. If we denote the result of the i-th test for a given e-mail as x_i, where $x_i = 1$ if the test succeeds and 0 otherwise, and we denote the weight of the i-th test as w_i, then the total score of an e-mail can be expressed as $\sum_{i=1}^{M} w_i x_i$, making use of the fact that w_i contributes to the sum only if $x_i = 1$, i.e., if the test succeeds for the e-mail [1]. Using t for the threshold above which an e-mail is classified as spam, the classifier had been written as $\sum_{i=1}^{M} w_i x_i > t$. Notice that the left-hand side of this inequality is linear in the x_i variables. Changing the inequality to an equality $\sum_{i=1}^{M} w_i x_i = t$ or $\mathbf{w} \cdot \mathbf{x} = t$, we obtain the *decision boundary*, separating spam from ham. The vector $\mathbf{w}$ is perpendicular to this plane. Figure 5 visualizes this for two variables. If there exists a linear decision boundary separating the two classes, we say that the data is *linearly separable*.

However, because linearly separable data does not uniquely define a decision boundary, we are now faced with a problem: which of the infinitely many decision boundaries should we choose? One natural option is to prefer large *margin* classifiers, where the margin of a linear classifier is the distance between the decision boundary and the closest instance. Support vector machines, developed at AT&T

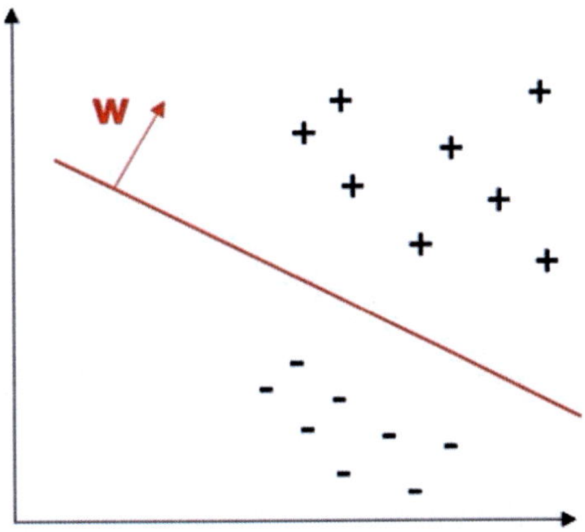

Fig. 5 An example of linear classification in two dimensions. The straight line separates the positives from the negatives and **w** points in the direction of the positives

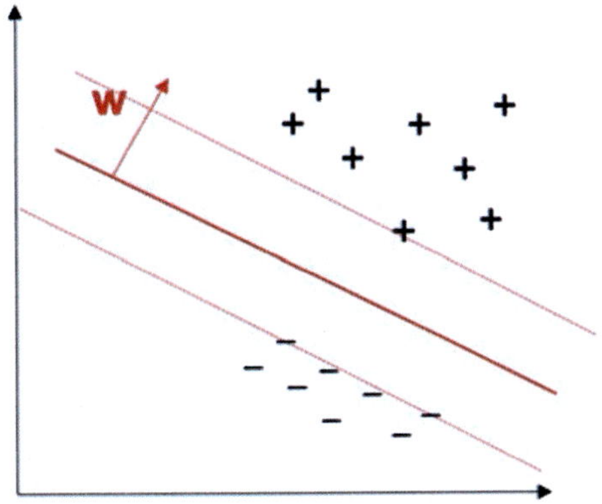

Fig. 6 The decision boundary learned by a support vector machine from the linearly separable data from Fig. 5. The decision boundary maximizes the margin, which is indicated by the dotted lines

Bell Laboratories by Vladimir Vapnik and colleagues [10], are a powerful kind of linear classifier that find a decision boundary whose margin is as large as possible.

For a given training set and decision boundary, let m^+ be the smallest margin of any positive, and m^- the smallest margin of any negative, then we want the sum of these to be as large as possible. This sum is independent of the decision threshold t, as long as we keep the nearest positives and negatives at the right sides of the decision boundary. Figures 6 and 7 [1] depict this graphically in a two-dimensional instance space.

The training examples nearest to the decision boundary are called **support vectors**: the decision boundary of a support vector machine is defined as a linear combination of the support vectors. The margin is thus defined as $\frac{m}{||\mathbf{w}||}$, where m is the distance between the decision boundary and the nearest training instances (at least one of each class) as measured along $\mathbf{w}$. Since we are free to rescale t, $||\mathbf{w}||$ and m, it is customary to choose $m = 1$. Maximizing the margin then corresponds to minimizing $||\mathbf{w}||$ or, more conveniently, $\frac{1}{2}||\mathbf{w}||^2$, provided that none of the training points fall inside the margin. This leads to a quadratic, constrained optimization problem:

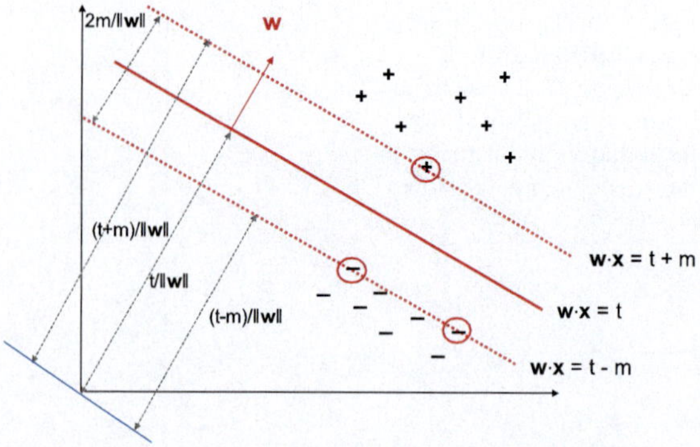

Fig. 7 The circled data points are the support vectors, which are the training examples nearest to the decision boundary

$$\mathbf{w}^*, t^* = \underset{\mathbf{w},t}{argmin}\frac{1}{2}||\mathbf{w}||^2 \text{ subject to } y_i(\mathbf{w}\cdot\mathbf{x}_i - t) \geq 1,\ 1 \leq i \leq n$$

with y_i taking value $+1$ or -1. By using the method of Lagrange multipliers α_i for each training example $i = 1..n$, we obtain that

$$\mathbf{w} = \sum_{i=1}^{n} \alpha_i y_i \mathbf{x}_i \quad (1)$$

The α_i are non-negative reals: if $\alpha_i = 0$ for a particular example $\mathbf{x}_i$, that example could be removed from the training set without affecting the learned decision boundary, meaning that $\alpha_i > 0$ only for the support vectors: the training examples nearest to the decision boundary. The dual optimization problem, it is entirely formulated in terms of the Lagrange multipliers:

$$\alpha_1^*, ..., \alpha_n^* = \underset{\alpha_1,...,\alpha_n}{argmax} -\frac{1}{2}\sum_{i=1}^{n}\sum_{j=1}^{n}\alpha_i\alpha_j y_i y_j \mathbf{x}_i\mathbf{x}_j + \sum_{i=1}^{n}\alpha_i \quad (2)$$

$$\text{subject to} \quad \alpha_i \geq 0,\ 1 \leq i \leq n \text{ and } \sum_{i=1}^{n}\alpha_i y_i = 0$$

Formula (2) shows that searching for the maximum-margin decision boundary is equivalent to searching for the support vectors: they are the training examples with non-zero Lagrange multipliers, and through (1), they completely determine the decision boundary. The majority of the α_i are 0, and correspond to the $\mathbf{x}_i$ that lie more than sufficiently away from the decision boundary and they have no effect on the plane. The instances that are not support vectors carry no information; even if any subset of them is removed, we would still get the same solution. The remaining data points are the support vectors, and they correspond to points that lie on the maximum margin (hyper-)planes in instance space. This property is central to the practical applicability of support vector machines: once the model is trained, a significant proportion of the data points can be discarded and only the support vectors retained.

An important property of support vector machines is that the determination of the model parameters corresponds to a convex optimization problem, and so any local solution is also a global optimum.

Soft Margin SVM If the data is not linearly separable, then the constraints $\mathbf{w} \cdot \mathbf{x}_i \geq t$ posed by the examples are not jointly satisfiable [1]. However, if we introduce slack variables ξ_i, one for each example, which allow some of them to be inside the margin (Fig. 8) or even at the wrong side of the decision boundary - we will call these margin errors - we can change the constraints to $\mathbf{w} \cdot \mathbf{x}_i - t \geq 1 - \xi_i$ and we result in the following soft margin optimization problem:

$$\mathbf{w}^*, t^*, \xi_i^* = \underset{\mathbf{w}, t, \xi_i}{argmin} \frac{1}{2} ||\mathbf{w}||^2 + C \sum_{i=1}^{n} \xi_i$$

$$\text{subject to} \quad y_i(\mathbf{w} \cdot \mathbf{x}_i - t) \geq 1 - \xi_i \text{ and } \xi_i \geq 0, 1 \leq i \leq n$$

C is a user-defined parameter: a high value of C means that margin errors incur a high penalty, while a low value permits more margin errors (possibly including misclassifications) in order to achieve a large margin.

5.1 *Training SVMs*

Although predictions for new inputs are made using only the support vectors, the training phase (i.e., the determination of the parameters α_i and t) makes use of the whole data set, and so it is important to have efficient algorithms for solving large quadratic optimization problems [11]. Direct solution using traditional techniques is

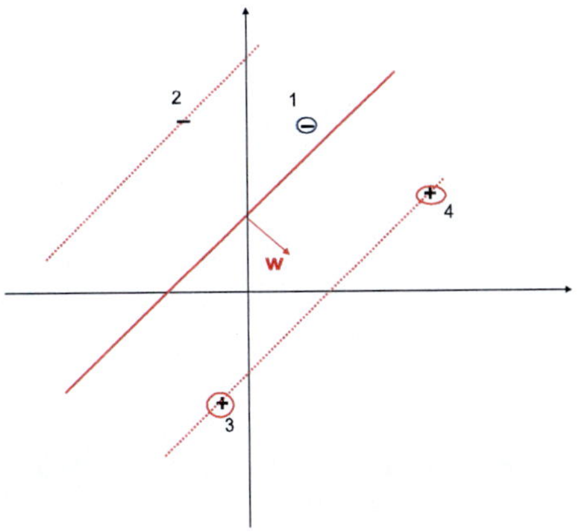

Fig. 8 A soft margin classifier: the negative training example no 1 is inside the margin

often infeasible due to the demanding computation and memory requirements, and so more practical approaches need to be found. The techniques of chunking [12], implemented using protected conjugate gradients [13], or Decomposition methods [14] try to solve a series of smaller quadratic programming problems. One of the most popular approaches to training support vector machines is called sequential minimal optimization, or SMO [15]. It takes the concept of chunking to the extreme limit and considers just two Lagrange multipliers at a time. In this case, the subproblem can be solved analytically, thereby avoiding numerical quadratic programming altogether. Heuristics are given for choosing the pair of Lagrange multipliers to be considered at each step. In practice, SMO is found to have a scaling with the number of data points that is somewhere between linear and quadratic depending on the particular application.

5.2 *Classification*

The support vector machine is fundamentally a two-class classifier.

For a two-class classification problem, in order to classify new data points using the trained model, we evaluate the sign of $y = \mathbf{w} \cdot \mathbf{x} - t$, where t represents a bias parameter. This can be expressed in terms of the parameters α_i and $\mathbf{x}$ by substituting for $\mathbf{w}$ using (1) to give $y(\mathbf{x}) = \sum_{i=1}^{n} \alpha_i y_i \mathbf{x}\mathbf{x}_i - t$. Having solved the quadratic programming problem and found a value for α_i, we can then determine the value of the threshold parameter t as a function of α_i, y_i, $\mathbf{x}_i$ considering only the number N_S of support vectors instead of the number of instances n.

Instead, the application of SVMs to problems involving $K > 2$ classes remains an open issue, as the approaches presented so far have limitations. In practice the 'one-versus-the-rest' approach [16] is the most widely used: we construct K separate SVMs, in which the k-th model $y_k(\mathbf{x})$ is trained using the data from class C_k as the positive examples and the data from the remaining $K - 1$ classes as the negative examples. If an input is assigned to multiple classes simultaneously, the prediction for new inputs $\mathbf{x}$ is done with $y(\mathbf{x}) = \underset{k}{max}\ y_k(\mathbf{x})$.

5.3 *Regression*

Support vectors can also be applied to regression scenarios, where we estimate a continuous-valued multivariate function. In this case we talk about *support vector regression (SVR)*. SVR is a generalization of SVM by means of the introduction of an ε-insensitive region around the function, called the ε-tube [10]. This tube reformulates the optimization problem to find the tube that best approximates the continuous-valued function, that is, finding a function $\hat{g}$ that has at most ε deviation from the actually obtained targets y_i for all the training data, and at the same time

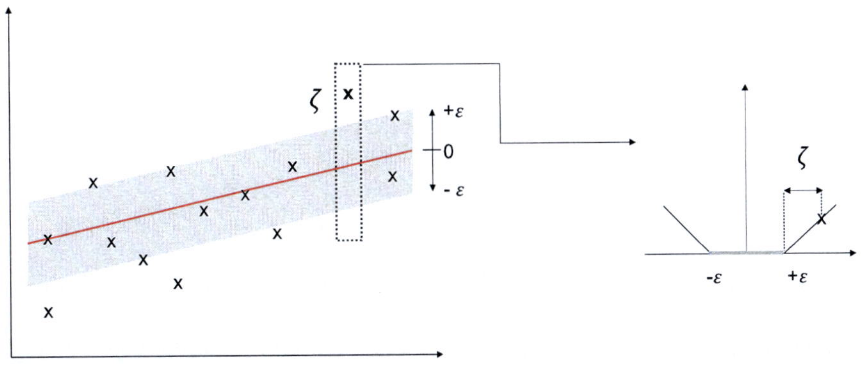

Fig. 9 A linear SVR

is as flat as possible. In other words, we do not care about errors as long as they are less than ε, but will not accept any deviation larger than this (Fig. 9, left) [17]. 'Flatness' in the case of $\mathbf{w} \cdot \mathbf{x}$ means that one seeks a small $\mathbf{w}$. One way to ensure this is to minimize the norm $||\mathbf{w}||^2$. We can write this problem as a convex optimization problem:

$$\begin{aligned} \text{minimize} \quad & \tfrac{1}{2}||\mathbf{w}||^2 \\ \text{subject to} \quad & y_i - \mathbf{w} \cdot \mathbf{x}_i + t \leq \varepsilon \\ & \mathbf{w} \cdot \mathbf{x}_i - t - y_i \leq \varepsilon \end{aligned}$$

More specifically, SVR is formulated as an optimization problem by first defining a convex ε-insensitive loss function to be minimized and finding the flattest tube that contains most of the training instances. Only the points outside the shaded region contribute to the cost, as the deviations are penalized in a linear fashion [18] (Fig. 9, right). Then, the convex optimization, which has a unique solution, is solved using appropriate numerical optimization algorithms. As in SVM, the support vectors in SVR are the most influential instances that affect the shape of the tube, and are points that lie on the boundary of the ε-tube or outside the tube.

6 Conclusions

This Chapter has been dedicated to introduce the main concepts of machine learning: we have seen how ML can build models from features for solving tasks involving data. We have seen how models can be predictive or descriptive, while learning can be supervised or unsupervised. We analyzed a few kinds of models, tree-structured and linear (specifically support vector machines), used for classification and regression. There are many successful applications of machine learning in a myriad of domains: there are commercially available systems for recognizing speech and handwriting. Retail companies analyze their past sales data to learn their customers' behavior to improve customer relationship management. Financial institutions analyze past transactions to predict customers' credit risks. Robots learn to optimize their behavior to complete a task using minimum resources. In bioinformatics, the huge amount of data can be analyzed and knowledge extracted using computers.

The entire next chapter will be dedicated to deep learning, a subfield of machine learning. Deep learning distinguishes itself from classical machine learning by the type of data that it works with and the methods with which it learns. Machine learning algorithms leverage structured, labeled data to make predictions, meaning that specific features are defined from the input data for the model and organized into tables. This does not necessarily mean that it does not use unstructured data; it just means that if it does, it generally goes through some pre-processing to organize it into a structured format.

Deep learning algorithms can ingest and process unstructured data, like text and images, and automate feature extraction, removing some of the dependency on human experts. For example, let's say that we have a set of photos of different pets, and we want to categorize them by "cat", "dog", etc. Deep learning algorithms can determine which features (e.g. ears) are most important to distinguish each animal from another. In machine learning, this hierarchy of features would be established manually by a human expert.

References

1. P. Flach, *Machine Learning: The Art and Science of Algorithms That Make Sense of Data* (Cambridge University Press, USA, 2012)
2. Y.S. Abu-Mostafa, M. Magdon-Ismail, H.T. Lin, *Learning From Data*. AMLBook (2012)
3. J.E. van Engelen, H. Hoos, A survey on semi-supervised learning. Mach. Learn. **109**, 373–440 (2019)
4. J.R. Quinlan, Induction of decision trees. Mach. Learn. **1**(1), 81–106 (1986). https://doi.org/10.1023/A:1022643204877
5. L. Breiman, J.H. Friedman, R.A. Olshen, C.J. Stone, *Classification and Regression Trees* (Wadsworth and Brooks, Monterey, CA, 1984)
6. J.R. Quinlan, *C4.5: Programs for Machine Learning* (Morgan Kaufmann Publishers Inc., San Francisco, CA, USA, 1993)
7. S.K. Murthy, S. Kasif, S. Salzberg, A system for induction of oblique decision trees. J. Artif. Int. Res. **2**(1), 1–32 (1994)

8. O.T. Yildiz, E. Alpaydin, Omnivariate decision trees. IEEE Trans. Neural Networks **12**(6), 1539–1546 (2001). https://doi.org/10.1109/72.963795
9. L. Breiman, Random forests. Machine Learning **45**(1), 5–32 (2001)
10. C. Cortes, V. Vapnik, Support-vector networks. Mach. Learn. **20**(3), 273–297 (1995). https://doi.org/10.1023/A:1022627411411
11. C. Bishop, *Pattern Recognition and Machine Learning* (Springer, Information Science and Statistics, 2016)
12. V. Vapnik, *Estimation of Dependences Based on Empirical Data: Springer Series in Statistics* (Springer-Verlag, Berlin, Heidelberg, 1982)
13. C.J.C. Burges, A tutorial on support vector machines for pattern recognition. Data Min. Knowl. Discov. **2**(2), 121–167 (1998). https://doi.org/10.1023/A:1009715923555
14. E. Osuna, R. Freund, F. Girosi, *Support Vector Machines: Training and Applications* (Tech. rep, Massachusetts Institute of Technology, USA, 1997)
15. J.C. Platt, Fast training of support vector machines using sequential minimal optimization, in *Advances in Kernel Methods: Support Vector Learning* (MIT Press, Cambridge, MA, USA, 1999), pp. 185–208
16. V.N. Vapnik, *Statistical Learning Theory* (Wiley-Interscience, 1998)
17. A. Smola, B. Schölkopf, A tutorial on support vector regression. Stat. Comput. **14**, 199–222 (2004)
18. M. Awad, R. Khanna, Support vector regression, in *Efficient Learning Machines* (Apress, Berkeley, CA, 2015), pp. 67–80

Neural Networks and Deep Learning Fundamentals

Riccardo Zese, Elena Bellodi, Michele Fraccaroli, Fabrizio Riguzzi, and Evelina Lamma

Abstract In the last decade, Neural Networks (NNs) have come to the fore as one of the most powerful and versatile approaches to many machine learning tasks. Deep Learning (DL), the latest incarnation of NNs, is nowadays applied in every scenario that needs models able to predict or classify data. From computer vision to speech-to-text, DL techniques are able to achieve super-human performance in many cases. This chapter is devoted to give a (not comprehensive) introduction to the field, describing the main branches and model architectures, in order to try to give a roadmap of this area to the reader.

1 A Brief History

In the last decades, the use of Neural Networks (NNs) has become one of the most effective approaches for solving classification and regression tasks. This is principally due to their capability of identifying and modeling complex correlations. They have been proposed for the first time in the 40s in order to try to achieve two main objectives: to study the functioning of the human brain by defining models able to simulate its neuro-physiological phenomena, and to use such models to extract

R. Zese (✉)
Department of Chemical, Pharmaceutical and Agricultural Sciences, University of Ferrara, Ferrara, Italy
e-mail: riccardo.zese@unife.it

E. Bellodi · M. Fraccaroli · E. Lamma
Department of Engineering, University of Ferrara, Ferrara, Italy
e-mail: elena.bellodi@unife.it

M. Fraccaroli
e-mail: michele.fraccaroli@unife.it

E. Lamma
e-mail: evelina.Lamma@unife.it

F. Riguzzi
Department of Mathematics and Computer Science, University of Ferrara, Ferrara, Italy
e-mail: fabrizio.riguzzi@unife.it

R. Micheloni and C. Zambelli (eds.), *Machine Learning and Non-volatile Memories*,
https://doi.org/10.1007/978-3-031-03841-9_2

the principles guiding human reasoning in terms of mathematical calculations, to develop artificial systems able to reason as a human but possibly faster and more efficiently.

The concept of NN, as the name explicitly says, was originally used to define networks simulating the neurons and their interactions in the human brain. The first theory was developed by McCulloch and Walter Pitts [31] and was very simple and effective. They defined the neuron as a computational unit that applies a function to the inputs implementing a binary classification. The inputs were multiplied by weights fixed apriori and static. In 1958, this first definition was at the basis of the implementation of the first model of neural network, called Perceptron [39], allowing the training of a single neuron. From the work of those years, besides the Perceptron, a second learning algorithm also emerged, which was a special case of the *Stochastic Gradient Descent* (SGD) technique, which is at the basis of most learning algorithms now. This learning algorithm was used to train the weights of the ADAptive LINear Element (ADALINE) model [48], used for linear regression. Basically, SGD needs an error function that returns, for a given output of the model, how far the output is from the label of the input. SGD takes the error function and computes the gradient of this error on the weights. In this way, it is possible to update the weights moving along the gradient in order to reduce the error. Performing these operations iteratively allows minimizing the error. Nowadays, this idea has been declined in many ways, considering for example also the second derivative or the derivative of the previous iterations to better guide the tuning of the weights. However, a single neuron is not effective for even the simplest classification and regression task because it implements a too simple function. For this reason, this idea was abandoned at that time.

In the second half of the eighties, Rumelhart et al. [40] used the back-propagation mechanism to train larger networks, giving new life to the study of the topic. The idea was simple but effective: using more neurons means combining more functions, defining a model able to represent more complex scenarios. Unfortunately, at that time, developing and experimenting with these ideas was difficult due to hardware limitations that made the training unfeasible with the increase of the complexity of the model. However, the idea of *deep networks* dates back to these years. Deep networks are networks with many layers of neurons, the internal layers are called hidden and the depth of the network is the number of layers. Moreover, the eighties saw the first theorization of Convolutional Neural Networks (CNN) by Yan LeCun [27], whose work resulted in the definition of the well-known LeNet5 model [26]. This model was one of the first effective CNN as it was able to achieve super-human results in the task of handwritten digit recognition.

Nowadays, the idea of NNs is that of extremely complex networks with many neurons organized in many layers. The concept of depth is stressed even further, partly thanks to the advances in hardware. Other important aspects that helped to increase the importance of these models are the increase of the size of the data and the availability of powerful (and in many cases user-friendly) systems and frameworks for the design, the implementation and the use of these models. Frameworks such as Tensorflow,[1] PyTorch[2] and Caffe[3] allow a better user friendliness and consequently a easier prototyping of models.

The classical definition of a NN, a set of neurons grouped in different layers where a neuron in a layer communicates with all the neurons of the next layer, is now usually called Fully Connected (Deep) Neural Network, Artificial Neural Network, Multilayer Perceptron, or Deep Feedforward Network [1, 13], and has been later extended in order to define more complex models.

Alongside this definition, new models have gained increasing importance in the field. On one hand, the already mentioned convolutional networks, mainly used in the field of computer vision, and on the other hand Recurrent Neural Networks [40], defined for input data in the form of sequences, such as written text. Their great innovation is that they allow the network to maintain a memory of previous data, enabling the management of such sequences.

The rest of this chapter is organized as follow. Section 2 discusses Multilayer Perceptrons, Sect. 3 introduces Convolutional Neural Networks, and Sect. 4 Recurrent Neural Networks. Section 5 discusses the problem of tuning the hyper-parameters to guide the training. Finally, Sect. 6 concludes the chapter.

2 Multilayer Perceptrons

The objective of a Multilayer Perceptron (MLP) is that of approximating a function $\hat{f} : \mathbb{R}^{N\times M} \rightarrow \mathbb{R}^K$ by means of other functions such that $\hat{f}(X) = f_n(f_{n-1}(\ldots (f_1(f_0(X)))))$, where X is a tensor[4] of size $N \times M$ representing the input. The corresponding model is a network of n layers, each representing the function f_i with i the number of the layer. Figure 1 shows an example of MLP with 4 layers. Layer 0 is the input layer, the leftmost one, then layers 1 and 2 are internal layers, also called *hidden layers*. Finally, the last layer is the output one, returning the results of the computation of the network. The output function is $\hat{Y} = \hat{f}(X) = f_3(f_2(f_1(f_0(X))))$, with $\hat{f} : \mathbb{R}^{3\times 1} \rightarrow \mathbb{R}^3$. This network is developed to solve a multiclass classification where each input x can be labelled with three different classes, associated with different neurons of the output layer. Before describing the overall flow, let us concentrate on a single neuron, e.g., neuron h_1 in Fig. 1. Figure 2 shows the computation performed in each single neuron. Basically, each neuron takes as input the output of each neurons of the previous layer, for the case of h_1, it takes $X = [x_1, x_2, x_3]^\top$ as input. Moreover, each neuron has another input value, the *bias*, a constant term set to 1 which represents background noise. Thus, toghether with the weight matrix W_1, containing one weight w_i^1 for each x_i, another weight b_1^1 is considered, associated to the bias.

[1] https://www.tensorflow.org/.

[2] https://pytorch.org/.

[3] https://caffe.berkeleyvision.org/.

[4] Generally, a tensor is a n-dimensional array with $n \in \mathbb{N}$.

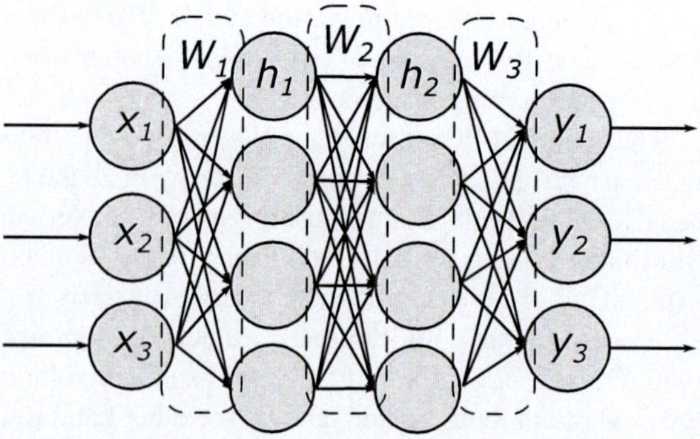

Fig. 1 Example of a MLP with four layers: the input, two hidden and the output layers. The input $X = [x_1, x_2, x_3]^\top$ represents a tensor of size 3×1. The output $Y = [y_1, y_2, y_3]^\top$ represents the 3 possible classes to be assigned to the input, one class for each value in Y

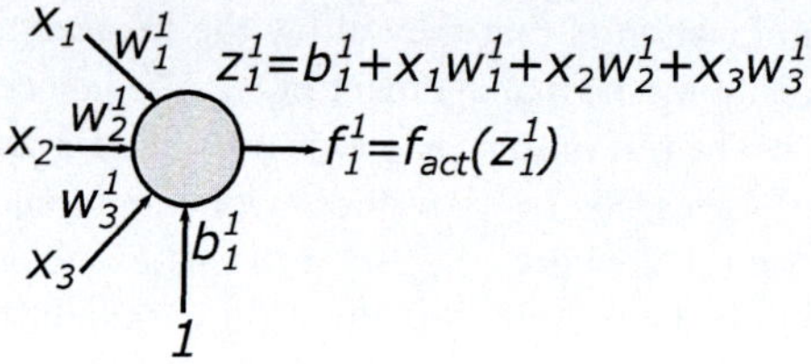

Fig. 2 Computation of neuron h_1 in Fig. 1

This weight is usually automatically added to the neuron by the frameworks used to model the MLP, thus it is not necessary to explicitly add it to the model, as shown in Fig. 1, where the bias terms are not represented.

In detail, first of all, the neuron computes the *network input*, i.e., the input it takes by the network. So, $z_1^1 = b_1^1 + \sum_{i=1}^{n} x_i w_i^1$ or, in matrix notation $z_1^1 = b_1^1 + X^\top W_1$, with $n = 3$ the size of X, i.e., the number of input from the previous layer. Then, z_1^1 is given to the *activation function* f_{act} that computes the output of the neuron, which is given as input to the neurons of the next layer. Thus, $f_1^1(X) = f_{act}(b_1^1 + X^\top W_1)$.

There are many possible activation functions, varying from hyperbolic tangent tanh to the sigmoid or to Rectified Linear Unit (ReLU) [24]. In particular, the last two functions are the most used. Given Z the input of a neuron, the sigmoid function $\sigma(Z) = \frac{1}{1+e^{-Z}}$, shown in Fig. 3a, is used as the activation function of the output layers in case of binary classification. It returns a value between 0 and 1, thus the output layer is designed to have one neuron returning the probability of the input to belong to the positive class (one of the two classes). On the other hand, the ReLU activation function, shown in Fig. 3b, is the most used for the neurons of the hidden layers. It was introduced to improve the performance in terms of training time and defined as $ReLU(H) = \max(0, Z)$.

Therefore, the overall flow is as follow. The input data is multiplied by weights matrix W_1 and given to the neurons of the first layer. Here, every neuron takes the result of this multiplication and applies the activation function, returning the output of the neuron. This value is intended as an indicator of the level of activation of the neuron, the higher the value, the most active that neuron.

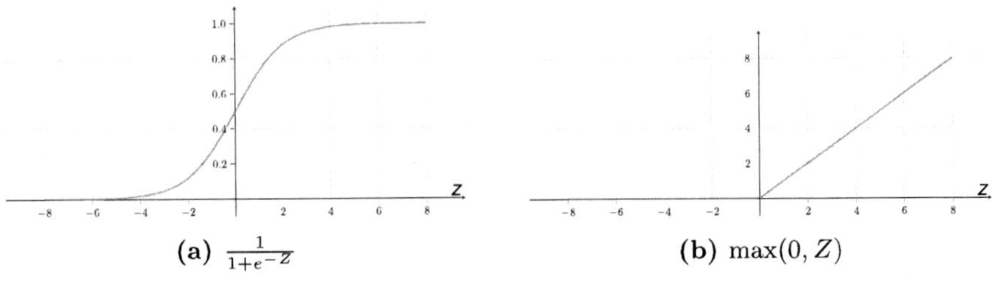

Fig. 3 Sigmoid **a** and ReLU **b** activation functions

This process is then repeated sequentially for the next layers until the output layer is reached. Thus, for example, the output of the neuron h_2 in Fig. 1 is $f_1^2(X) = f_{act}(b_1^2 + Z_1^\top W_2)$ where b_1^2 is the bias weight of neuron h_2, and Z_1 is the tensor containing the output of the neurons of the previous layer.

The last layer will output a probability distribution among the classes considered in the data in the case of classification, or a real value predicting the output label of the example in the case of regression.

During learning, this output distribution is used to compute a loss measure representing how far the predicted output of the network is from the actual label of the input examples. The gradient of this loss measure is calculated w.r.t. the weights. These are then updated by taking a step in the opposite of the direction given by the gradient by means of the gradient descent algorithm or one of its evolutions. This approach is called *back-propagation* and it is at the basis of the training of each neural network, with some variants due to the different architectures adopted.

The need of calculating the gradient guided the search for good activation functions for all the layers. What one would like to have is a function easy to derivate, able to maintain the output values of each neuron in a range that is functional for the training and, at the same time, that does not reduce the information passing through the neuron. Values in the range [0, 1] are the easiest to manage because they avoid an explosion of the weights' values. The counterpart is that multiplying many values lesser than 1 makes the output close to 0, making its gradient close to 0 as well. This problem is called *vanishing gradient*. To try to avoid this problem, the choice of the value of the update step for the weights must be wisely chosen and, in many cases, it must be changed during the training as well.

3 Convolutional Neural Networks

When the input data presents some spatial structure, i.e., each value of the input is connected in some way with other values, MLP may have problems to represent these spatial relations, that usually connect smaller portions of the input example and may appear in different positions and numbers. MLPs need a neuron for each value of the input example, which is a tensor, thus they may require too many neurons to effectively handle the problem. Input data with spatial structure are, e.g. images, where each pixel is connected with the pixels surrounding it, or time series, where each value is connected with the previous and the following value in the series. If we consider a picture of a person, the set of pixels representing an eye are related and the same relation appears in two different places inside the picture. Moreover, the eye is related to the face of the subject but not, for example, to a car passing behind the person in the background. In these cases, if a MLP is considered, the input layer should have a neuron for each value of the input, i.e., one neuron for each pixel in case of a greyscale image or three neurons for each pixel in case of an RGB picture (one for each channel). This may force the network to have an extremely large number of weights to tune. Consider, for example, a greyscale image with a resolution of 256×256 pixels. The input layer has 65,536 neurons. If the first hidden layer has 100 neurons, which is usually a number too small to be effective given the input size, the network needs 65,536·100=6,553,600 weights to connect the two layers. This number will increase even further adding more layers, resulting in a hard to train network.

Convolutional Neural Networks (CNNs) [27] have come to the fore to solve problems where data presents spatial and/or topological structure as in the previous example. CNNs are built using convolutional layers, performing convolutions on the data they receive and extracting *features* from it, typically followed by some fully connected layers to carry out the classification by considering only the extracted features instead of the entire input data. Thus, considering the previous example, the convolutional part checks which features are present in the picture, e.g., if eyes or a mouth or wheels are present. Then, the fully connected part considers the set of features identified in the image and decides which label to assign.

The convolution operation is the integration of two functions x and w, and it is denoted as $x * w$. Function x is called input while function w is the kernel. The result is called feature map, or simply feature. Convolution is defined as

$$(x * w)(t) = \int_{-\infty}^{\infty} x(\tau)w(t - \tau)d\tau$$

or, in the discrete case

$$(x * w)(t) = \sum_{\tau=-\infty}^{\infty} x(\tau)w(t - \tau)$$

In practical scenarios, the input and the kernel functions are tensors, i.e., multidimensional arrays, such as a 2D grid of pixels for an image or a 1D vector for a time series. The kernel scrolls all the input, moving in the directions given by the size of the input itself (in one direction in the case of 1D input, in two directions in the case of 2D input, etc.). Good surveys on the application of CNNs to images and time series are respectively [20, 36].

The result of a convolution operation represents the features extracted from the input. The advantage of using CNNs is that the kernel is used on the entire input. The only weights to train are the values of the kernel, therefore, if we consider a single kernel of size 5×5, the convolutional layer has only 25 weights to train irrespective of the size of the input of the layer. Even if the input has size 256×256, the number of weights to train will be always 25.

In practice, every convolutional layer applies many kernels at a time, extracting many features from the same image, one for each kernel. For example, one of the first CNN, LeNet5 [26], was defined to train 10 kernels of size 28×28 in the first convolutional layer. Later, Simonyan and Zisserman [42] showed that it is possible to further reduce the number of weights by maintaining the same receptive field by reducing the size of the kernel and adding more convolutional layers, as shown in Fig. 4.

The receptive field is the number of values of the input affecting a single value of the output of a convolution operation. Considering a single convolutional layer having 5×5 kernels, the receptive field of the output of this layer is 25 (Fig. 4a). Consider now replacing this single layer with two convolutional layers having 3 kernels each, as depicted in Fig. 4b. After the second convolutional layer, every value of the output depends on 25 values of the input given to the first convolutional layer, i.e., the receptive field of each value of the output of the second layer is a square 3×3, so 9 values, of the output of the first layer. In the output of the first layer, each value has a receptive field of 3×3 values of the input. By considering these two layers as a single black-box, the receptive field of the output combines those of the two layers, becoming, as said before, a square of 5×5 values of the input. This is true if we consider a stride equal to 1. The stride indicates by how many pixels the kernel moves in a certain direction to calculate the feature map. The advantage of using more layers is that, while with a single 5×5 convolutional layer there are 25 weights to train, with two 3×3 convolutional layers there are $2 \cdot 9 = 18$ weights to train. If we consider a single layer having a 7×7 kernel and replace it with three layers having 3×3 kernel and using stride 1, the same receptive field is achieved with $3 \cdot 9 = 27$ weights instead of $7 \cdot 7 = 49$. Since every convolutional layer contains several different kernels, the gain in terms of number of weights to train increases fast. By exploiting this approach, it is possible to extract hundreds of features by maintaining the number of weights to train feasible.

However, with the increase of layers, the problem of vanishing gradient will pop up. For this reason, He et al. [16] introduced ResNet where the configuration of layers combines the output of sets of convolutional layers with the input of the first layer in the sets creating a short-circuit between the input and the output of the set of layers, as shown in Fig. 5. This configuration of layers is called *Residual Block* and presents an output function that is $H(X) = F(X) + X$, where $F(X)$ is the output of the set of convolutional layers. Moreover, the increase of complexity of a layer in terms of operations performed also implies an increase of the number of weights. To reduce their number, a possibility is to resort to 1×1 kernels, as in GoogLeNet [46]. Basically, 1×1 convolution is used to reduce the number of features computed by the previous convolutional layer by combining them in a meaningful way. This combi-

Fig. 4 Comparison between 1 convolution operation with kernel 5 × 5 (a) and 2 convolution operations with kernel 3 × 3 (b)

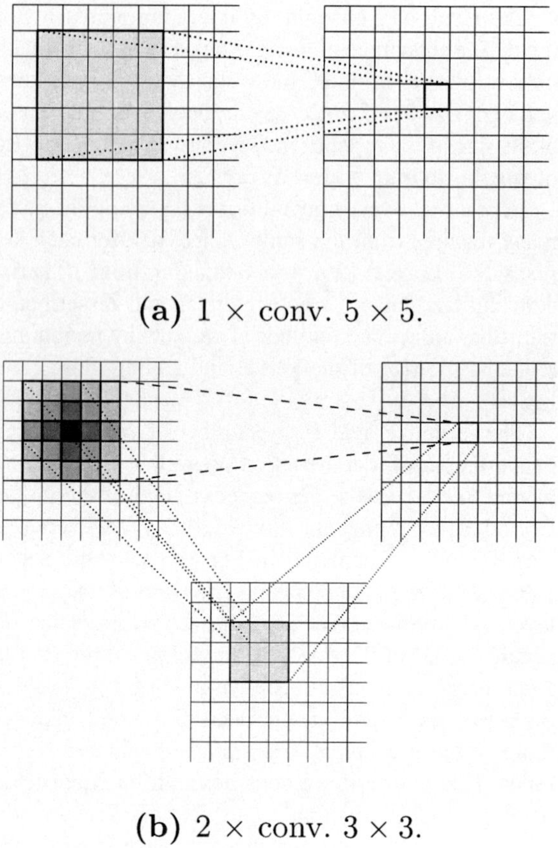

Fig. 5 The architecture of a Residual Block as defined in ResNet

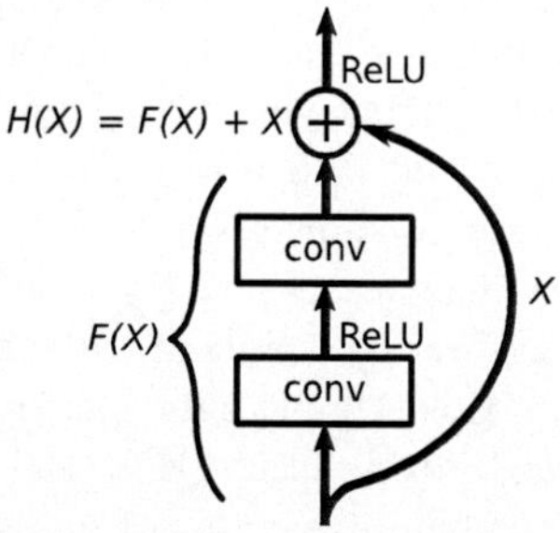

nation is learned automatically during the training phase to maintain the information of the features taken as input by the 1×1 convolutional layer while reducing their number. GoogLeNet also introduced the concept of *inception*, i.e., the design of a good local convolutional network architecture, usually with parallel branches, and its use as a new layer.

In the last years, many other different architectures have been presented. Most of them combine and extend the ideas presented for GoogLeNet and ResNet by including MLP networks inside layers. An example is the well-known Network-in-Network model [28], combining the idea of residual blocks and inception [49], creating fractal architectures [25], where the model is no more defined by means of layers but by means of a fractal function which is defined recursively, or heavily using 1×1 convolution [19] or short-circuits [18].

Another important operation used by CNNs to reduce feature maps is *pooling*, also called sub-sampling, which usually follows convolution operations. This is used to reduce the size of a feature map by summarizing its information. This summarization can be performed by, e.g., averaging neighbour values or extracting from them the maximum value. The use of convolution and pooling operations allows the extraction of the important features contained in the input data X, representing them by smaller tensors. These features can be used to help classification, or the recognition of certain patterns contained in the data, irrespectively of their position or scale.

Indeed, CNNs' final layers are usually classical fully connected layers that perform classification on the features extracted by previous layers. Therefore, roughly speaking, a CNN can be divided into two subnetworks, where the first subnetwork processes the input data by means of convolution operations. The output of these operations is then passed to the second subnetwork, which is a fully connected network used to classify the input data. Figure 6 shows the architecture of LeNet5 (Fig. 6a) [26] and of ResNet (Fig. 6b) [16]. From this figure it is possible to see that while the overall architecture has remained the same, i.e., a convolutional part that feeds a set of features to a fully connected part, the depth has increased significantly, going from 6 layers of LeNet5 to 152 layers of ResNet.

Considering images as input, besides the classification task, CNNs are also used to solve more specialized tasks, for example, semantic and instance segmentation, or object detection. The first task consists of identifying which part of the image each pixel belongs to. For example, consider a picture having two dogs in the foreground and a grassland with sky as background, semantic segmentation's objective is that of "classifying" each pixel as sky, grass and dog, while instance segmentation adds the capability of discriminating among the two dogs. Thus, the difference is that semantic segmentation does not discriminate between different subjects, the pixels representing the two dogs are all classified as "dog", while in instance segmentation the pixels representing the first dog are kept separated from those representing the second dog. This task is performed by superimposing a mask on the initial image that colours each pixel depending on how it is classified. This task is usually performed by applying convolution and deconvolution to the image so that the fully connected part of the network is replaced by a sequence of deconvolutional layers [30, 32, 41]. In particular, deconvolution is the dual of convolution, i.e., instead of extracting features from an image, reducing its size, it is aimed at recreating an image starting from a set of features, also possibly increasing its size.

Object detection allows to locate and surround with a box the objects represented in the image. Therefore, given an image, the objective is to identify which objects the image contains, classify them and spatially locate them in the image. This task

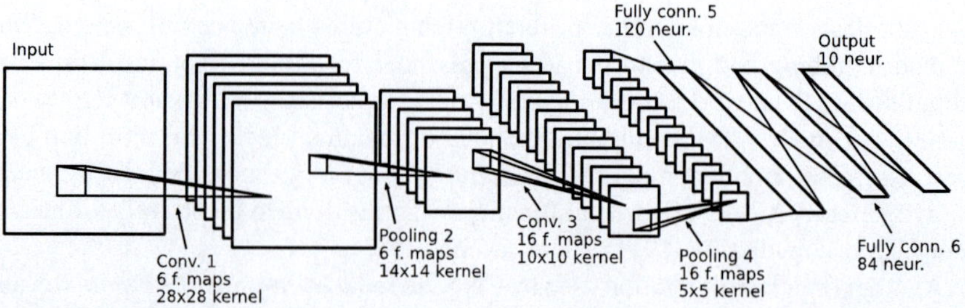

(a) LeNet5 [26] executes convolution two times (layers Conv. 1, and Conv. 3), and two pooling operations (Pooling 2, and Pooling 4). Finally, two fully connected layers (5 and 6).

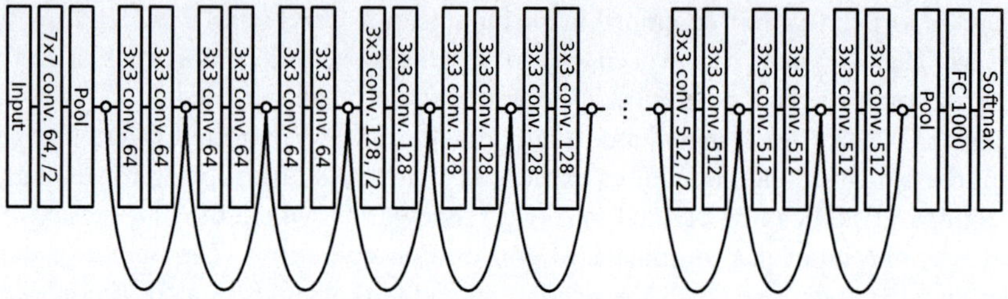

(b) ResNet [16] starts with a convolution and a pooling layer. Then, 148 convolutional layers divided into 74 residual blocks composed of two layers. At the end, one fully connected layer (FC) connects the extracted features to the output layer that uses as activation function the *softmax*, which acts similarly to the sigmoid for multi-class classification.

Fig. 6 The architecture of LeNet5 (**a**) and ResNet (**b**)

is performed by defining and training the fully connected layers at the end of the model so that they also output the coordinates of the box highlighting each object. This approach can be used also to estimate the pose of the subject of the image [47], for example to identify if a person is sitting or walking. Object detection and pose estimation can be done by iteratively selecting portions of the image randomly sampled and using this portion as input of the CNN. In this way, the model can extract the information contained in each portion and thus in the entire image [12].

4 Recurrent Neural Networks

If we must process sequential data, the architectures seen before are not the best choice. Sequential data may have different length and be very long, however the model needs to be able to analyse the sequence as a whole. MLPs and CNNs take as input data of fixed size and are designed to handle different characteristics in the data, such as grid-like topology for the CNNs. When sequential data needs models having a memory of what they have seen previously in the sequence, *Recurrent Neural Networks* (RNNs) [40] may come in handy. An example of an application of the RNNs can be text translation, where given a sentence, the system has to produce a different sentence, which is the translation of the input in a different language. The vanilla RNN is composed of a single neuron that takes as input a single value of the sequence at a time and a value representing its previous state, i.e., the state, usually the output or a function of the output, obtained considering the previous values of the sequence. In a sequence, every individual value represents the value of the sequence at a certain time step t, therefore, the output of a RNN is $h_t(x_t) = f(h_{t-1}, x_t)$ for $t \geq 0$, h_t is the output of the recurrent layer at time t, f the function computed by the layer, and x_t the input at time t. The function f can vary a lot and different types of functions have been defined to manage the process memory in different ways, by replacing the single neuron of the vanilla version with sub-networks containing more neurons. Each version of f uses weights that are learned during the training phase. An important thing to note is that the network is composed by a single neuron/sub-network applying function f, therefore, the weights are tuned by considering the whole sequences, while the network needs less memory than the other types of network discussed in this chapter.

This simple architecture can be used for different purposes, given an input sequence the model could return as output a different sequence or a single value, or else, the model could take a single value as input instead of a sequence and use only its previous state to create a sequence from scratch. This can be easily done by deciding when to return an output, either at each time step, at the end of the input sequence, or every intermediate solution.

The training method used for this type of network is called *back-propagation through time* and in its basic definition is computed for each time step. The difference is that in this case, at each time step, the number of times the weights of the network are considered increases. In fact, at time 1 the output considers the weights of the neuron only once, at time 2 the output is the output of time 1 multiplied by the weights, so they are considered twice, and this holds for every time step until the end of the sequence.

We can also define *Bidirectional-RNN*, where for each time step the output depends also on the future values of the sequence, thus the whole sequence must be read forward and backward in order to return the output. It is worth noting that the possibility of reading the sequence bidirectionally opens to the possibility of taking as input not only sequences but also input that have more dimensions, such as images. This can be done by considering different directions for each dimension, e.g., in the case of an image the network should consider 4 directions.

With the combination of RNNs and CNNs it is also possible to solve the image captioning problem, i.e., given an image, generate a caption describing the content of the image [22]. This can be done by exploiting a CNN trained for solving object recognition task. This network returns a set of labels telling what the image contains. A RNN is then fed with these labels to generate a caption, i.e., a sequence of words, for the image.

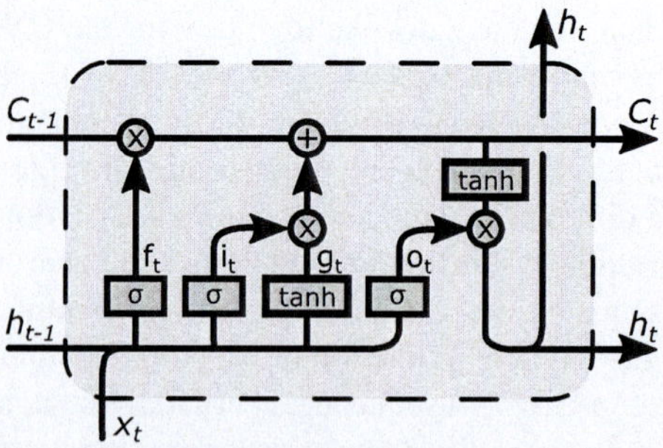

Fig. 7 Long Short Term Memory architecture

The main problem of vanilla RNNs is that the network tends to forget the most distant parts of the sequence. Thus, as seen before, more complex architectures have been proposed to improve the memory of the network. One of the most important architecture is called *Long Short Term Memory* RNN (LSTM) [17]. LSTMs replace the single neuron of the vanilla RNN with four gates (basically four neurons) that, for each time step, take as input the current value of the input sequence, and the previous output and state, as shown in Fig. 7. One gate decides which parts of the previous state C_{t-1} to remember by checking the previous output h_{t-1}, another gate computes which parts of the current input x_t to remember, a third gate decides how these two parts are combined together to compute the current state C_t and so to manage what to keep in memory, and the fourth gate computes the current output h_t.

As can be seen in Fig. 7, an LSTM is a network containing several MLP neurons, called gates. To compute the output and the state of the LSTM, it is necessary to compute the output of each gate.

The output of gate f at time t, called *forget* gate, telling what to forget from the previous state is

$$f_t = \sigma(W_f \cdot [h_{t-1}, x_t] + b_f)$$

where $\sigma(\cdot)$ is the sigmoid function, W_f and b_f are the weights matrix and the bias weight of gate f,and $[h_{t-1}, x_t]$ is the concatenation of the two tensors h_{t-1} and x_t.

The output of gate i at time t, called *input* gate and telling which values to update, is

$$i_t = \sigma(W_i \cdot [h_{t-1}, x_t] + b_i)$$

where W_i and b_i are the weights matrix and the bias weight of gate i.

The output of gate g at time t, called *gate* gate and creating a vector of new candidate values, is

$$g_t = \tanh(W_g \cdot [h_{t-1}, x_t] + b_g)$$

where $\tanh(\cdot)$ is the hyperbolic tangent, W_g and b_g are the weights matrix and the bias weight of gate g.

Now, it is possible to compute C_t as

$$C_t = f_t \times C_{t-1} + i_t \times g_t$$

where $\times$ is the element-wise multiplication.

To find the output h_t, the output of the gate o, called *output* gate, must be computed as:

$$o_t = \sigma(W_o \cdot [h_{t-1}, x_t] + b_o)$$

where W_o and b_o are the weights matrix and the bias weight of gate o. Finally, the output of the LSTM at time t can be computed as

$$h_t = o_t \times \tanh(C_t)$$

Starting from LSTM, a plethora of new models have been proposed. Some of them combine or replace gates to simplify the training, such as the well-known *Gated Recurrent Unit* (GRU) [7], or introduce *spyholes* [11] giving also the previous state as input to (some) of the four gates of a standard LSTM.

The idea of adding sub-networks has also been applied in the case of RNNs [14, 33]. For example, the current state may be passed as input to an MLP whose output is considered as the previous state from the recurrent part [14].

A generalization of RNNs are the *Recursive Neural Networks* [5, 35]. In this case, the model presents a tree-like structure, as shown in Fig. 8. The main advantage is that the number of operations necessary to compute the output given an input sequence is reduced to be logarithmic in the length of the sequence. The drawback is that to achieve the best results the structure of the tree should be tailored to the input. This structure is usually computed by analysing the task. For example, for natural language processing, one can exploit some parsers to get a parse tree to be transformed into the recursive network [44, 45], or define a learner able to automatically create the structure of the tree [5]. However, given the difficulties to correctly identify the best architecture, recursive neural networks are rarely used.

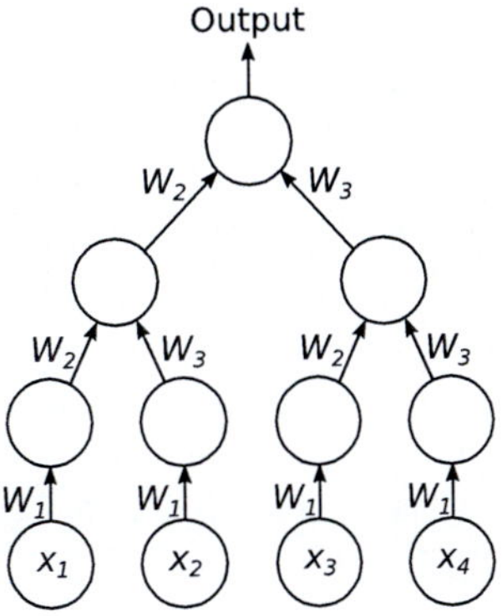

Fig. 8 Recursive neural network

5 Hyper-parameter Optimization

An important aspect to consider for each machine learning approach is the choice of good hyper-parameters. These can be the number of neurons in each layer, the number of layers, the size of the kernels or the number of iterations to perform during training. As seen in the previous sections, with the increasing sophistication of Deep Learning systems, the hyper-parameters and the possible configurations of the network architectures become more and more complex. Furthermore, due to the pervasiveness of Deep Learning systems, the tuning process of the hyper-parameters and the choice of the best neural architecture needs to be addressed even by non-experts.

In the automation of Deep Learning, we can distinguish two families of algorithms: Hyper-Parameters Optimization (HPO) and Neural Architecture Search (NAS) algorithms [9]. With HPO algorithms we modify only the hyper-parameters of the models while with NAS we act only on the architecture of the neural networks.

The four main HPO approaches are Grid Search, Random Search, Bayesian Optimization and with Genetic algorithm. Grid search [51] performs exhaustive search on the specified hyper-parameters space. This algorithm performs a new indipendent training session for each combination of the hyper-parameters. This algorithm ensures that the optimal configuration is found as long as sufficient resources are provided. This method is guided by a performance metric that decides which of the tested configurations is the best, measuring the performance of the neural model in the training or validation phase. However, due to the fact that the computational resources increase exponentially with the number of hyper-parameters to set, Grid Search suffers from the *curse of dimensionality* [51].

Random Search [3] searches randomly in the user-defined hyper-parameters space. Random search leads to better results than Grid Search, especially when only a small number of hyper-parameters affect the final performance of the learning algorithm [3]. Unlike Grid Search, this algorithm is not guaranteed to achieve the optimum, but may require less computation time while finding a reasonably good model in most cases [3] because the maximum computation time is set before starting the search. With Random Search, it is also possible to include prior knowledge by specifying the distribution from which to sample the hyper-parameter values.

Bayesian Optimization aims to find the global optimum with the minimum number of trials. It is a probabilistic model-based approach for optimizing objective functions which are very expensive or slow to evaluate [8]. Bayesian Optimization builds a probabilistic model (called *surrogate model*) of the objective function and quantifies the uncertainty in this surrogate using a regression model. Then, it uses an acquisition function to decide where to sample the next set of hyper-parameters values [10]. Its effectiveness in optimizing the hyper-parameters of NNs derives from the fact that it limits the number of training sessions by spending more time choosing the next set of hyper-parameters to try. In the literature, there are many examples of application of this kind of HPO to NNs [4, 23, 43].

A genetic algorithm (GA) is a metaheuristic based on Charles Darwin's evolution theory [50] frequently used to generate high-quality solutions to optimization problems. This algorithm tries to imitate the process of natural selection where the fittest individuals are selected for reproduction to produce the population of the next generation. Evolution starts with a randomly generated population of individuals (each individual is a solution to the optimization problem). One of the key points of GA is the *fitness function*. The fitness function determines the *fitness score* of each individual. Fitness score represents the probability that an individual will be selected for reproduction. At each iteration of the algorithm, with the *selection* phase, the fittest individuals are selected so that they pass on their genes to the next generation. Individuals with high fitness score have more chance to be selected for reproduction. Through the *crossover* phase (also called *recombination*), for each new solution to be produces, a pair of *parent* solution of the actual generation are selected and their genetic information are combined to create new *child* solution. In addition to the crossover, to maintain the genetic diversity, there is also the *mutation*. Mutation alters one or more gene values in an individual from its initial state and occurs during the creation of the new population, according to a user-definable mutation probability. The algorithm terminates where the population has converged (GA does not produce new population that are significantly different from the previous generation).

Then, in the case of DL hyper-parameters optimization, the Genetic algorithm starts with an initial population of N DL models with some predefined hyper-parameters. Then, we can calculate the accuracy (or loss) of the models and uses that as a fitness score. Finally, we can generate a new offspring of DL models. This method is slow (at each iteration, new neural networks are generated which need to be trained) and not guaranteed to find the optimal solution.

NAS is a technique for automating the design of NNs architectures strictly correlated to HPO. NAS methods have already been shown to be capable of overcoming manually designed architectures [37, 52]. The three main elements on which NAS are based are: *Search Space*, *Search Strategy* and *Performance Estimation Strategy* as can be seen in Fig. 9. Search Space refers to all possible architectures that can be generated during the optimization process. Search Strategy refers to the methods for exploring all possible architectures that can be generated by NAS. Performance Estimation Strategy are the methods for measuring the quality of the generated NN [9]. There are different search strategies that can be used to explore the search space of neural architectures. These strategies include: random search, Bayesian Optimiza-

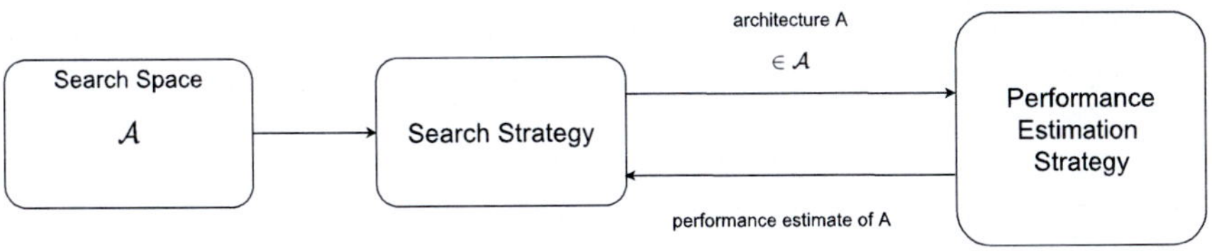

Fig. 9 Components of NAS with their interactions

tion [21], reinforcement learning [34], gradient-based methods [29] and evolutionary algorithms [9, 38]. NAS approaches can be categorized in two groups: classical NAS and one-shot NAS [2]. The first group follows the traditional search approach also used by Grid Search, where each generated NN is trained independently. One-shot NAS algorithms use weight sharing among models in the search space to train a super-net and use this to select better models. This type of algorithms reduces computation resources compared to the classical NAS algorithm. Therefore, a super-net is a single large network that contains every possible operation in the search space. All possible network architecture in the super-net can be considered as a sub-net with shared weights between common edges. Then, rather than training thousands of separate models from scratch like in the classical NAS, one can train a single large network (super-net) capable of emulating any architecture in the search space. Once the super-net is trained, it is used for evaluating the performance of many different architectures sampled at random by zeroing out or removing some operations.

The state-of-the-art of one-shot NAS are: Efficient Neural Architecture Search (ENAS) [34], Differentiable Architecture Search (DARTS) [29], Single Path One-Shot (SPOS) [15] and ProxylessNAS [6]. Nowadays, different software libraries implement this kind of algorithms. We can cite Autokeras[5] [21] and Neural Network Intelligence (NNI).[6] Autokeras implements one-shot NAS with Bayesian Optimization-like search strategy and NNI implements both HPO algorithms and classical and one-shot NAS algorithms.

6 Conclusions

This chapter illustrates the main concepts of the Deep Learning field. In particular, it discusses the most important architectures, i.e., Multilayer Perceptrons (MLPs), Convolutional Neural Networks (CNNs) and Recurrent Neural Networks (RNNs). While MLPs are a necessary component for performing classification or regression, the other two types are more tailored to specific input types.

CNNs are extremely useful when dealing with input that shows a grid-like topology, such as images and, to some extent, time series. These types of input data are multi-dimensional tensors where each value is connected with the neighbouring values. CNNs owe their success to their capability of extracting features from the input data.

RNNs are well suited for sequential data, such as sentences. Indeed, this kind of architecture is often used for natural language processing, both for reading sentences as input, and generating sentences as output. The main feature of RNNs, absent in CNNs and MLPs, is that they are designed to keep memory of the (most important parts of the) whole input sequence. Therefore, unlike CNNs, RNNs can consider the sequence as a whole instead of considering only a bunch of neighbouring values.

---

[5] https://autokeras.com/.

[6] https://www.microsoft.com/en-us/research/project/neural-network-intelligence/.

All these architectures can be combined to create powerful tools. For example, to classify images given their content, a CNN needs to send its output to a MLP, which will use the extracted features to perform the classification. Another interesting combination is between CNNs and RNNs to label images. In this case, the extracted features are given as input to a RNN to compose a sentence describing the content of the image.

The architecture of the networks and the algorithms used to train the networks depend on many hyper-parameters, that must be optimized to obtain good results. Since the number of these hyper-parameters can be high, we need mechanisms to automatically tune their values. For this reason, in this chapter we have surveyed the relevant literature about hyper-parameters' optimization.

References

1. C.C. Aggarwal, *Neural Networks and Deep Learning—A Textbook* (Springer, 2018)
2. G. Bender, P. Kindermans, B. Zoph, V. Vasudevan, Q.V. Le, Understanding and simplifying one-shot architecture search, in *Proceedings of the 35th International Conference on Machine Learning, ICML 2018*, Stockholmsmässan, Stockholm, Sweden, 10–15 July 2018, eds. by J.G. Dy, A. Krause. Proceedings of Machine Learning Research, vol. 80, pp. 549–558. (PMLR, 2018), http://proceedings.mlr.press/v80/bender18a.html
3. J. Bergstra, Y. Bengio, Random search for hyper-parameter optimization. J. Mach. Learn. Res. **13**, 281–305 (2012). http://dl.acm.org/citation.cfm?id=2188395
4. H. Bertrand, R. Ardon, M. Perrot, I. Bloch, Hyperparameter optimization of deep neural networks: combining hyperband with bayesian model selection, in *Conférence sur l'Apprentissage Automatique* (2017)
5. L. Bottou, From machine learning to machine reasoning—an essay. Mach. Learn. **94**(2), 133–149 (2014)
6. H. Cai, L. Zhu, S. Han, Proxylessnas: direct neural architecture search on target task and hardware. arXiv preprint arXiv:1812.00332 (2018)
7. K. Cho, B. van Merrienboer, D. Bahdanau, Y. Bengio, On the properties of neural machine translation: encoder-decoder approaches, in *Proceedings of SSST@EMNLP 2014, Eighth Workshop on Syntax, Semantics and Structure in Statistical Translation*, eds by D. Wu, M. Carpuat, X. Carreras, E.M. Vecchi (Doha, Qatar, 2014), pp. 103–111. Association for Computational Linguistics (2014). https://doi.org/10.3115/v1/W14-4012, https://www.aclweb.org/anthology/W14-4012/
8. I. Dewancker, M. McCourt, S. Clark, Bayesian optimization primer (2015), https://app.sigopt.com/static/pdf/SigOpt_Bayesian_Optimization_Primer.pdf
9. T. Elsken, J.H. Metzen, F. Hutter, Neural architecture search: a survey. arXiv preprint arXiv:1808.05377 (2018)
10. P.I. Frazier, A tutorial on bayesian optimization. arXiv preprint arXiv:1807.02811 (2018)
11. F.A. Gers, J. Schmidhuber, Recurrent nets that time and count, in *Proceedings of the IEEE-INNS-ENNS International Joint Conference on Neural Networks, IJCNN 2000, Neural Computing: New Challenges and Perspectives for the New Millennium*, Vol. 3, pp. 189–194. Como, Italy, 24–27 July 2000. IEEE Computer Society (2000). https://doi.org/10.1109/IJCNN.2000.861302
12. R.B. Girshick, J. Donahue, T. Darrell, J. Malik, Rich feature hierarchies for accurate object detection and semantic segmentation, in *2014 IEEE Conference on Computer Vision and Pattern Recognition*, CVPR 2014, pp. 580–587. Columbus, OH, USA, 23–28 June 2014. IEEE Computer Society (2014). https://doi.org/10.1109/CVPR.2014.81

13. I. Goodfellow, Y. Bengio, A. Courville, *Deep Learning*, vol. 1 (MIT Press, 2016)
14. A. Graves, A. Mohamed, G.E. Hinton, Speech recognition with deep recurrent neural networks, in *IEEE International Conference on Acoustics, Speech and Signal Processing, ICASSP 2013*, pp. 6645–6649. Vancouver, BC, Canada, 26–31 May 2013. IEEE Press (2013). https://doi.org/10.1109/ICASSP.2013.6638947
15. Z. Guo, X. Zhang, H. Mu, W. Heng, Z. Liu, Y. Wei, J. Sun, Single path one-shot neural architecture search with uniform sampling. arXiv preprint arXiv:1904.00420 (2019)
16. K. He, X. Zhang, S. Ren, J. Sun, Deep residual learning for image recognition, in *2016 IEEE Conference on Computer Vision and Pattern Recognition, CVPR 2016*, pp. 770–778. Las Vegas, NV, USA, June 27–30 2016. IEEE Computer Society (2016). https://doi.org/10.1109/CVPR.2016.90
17. S. Hochreiter, J. Schmidhuber, Long short-term memory. Neural Comput. **9**(8), 1735–1780 (1997)
18. G. Huang, Z. Liu, L. van der Maaten, K.Q. Weinberger, Densely connected convolutional networks, in *2017 IEEE Conference on Computer Vision and Pattern Recognition, CVPR 2017*, pp. 2261–2269. Honolulu, HI, USA, 21–26 July 2017. IEEE Computer Society (2017). https://doi.org/10.1109/CVPR.2017.243
19. F.N. Iandola, M.W. Moskewicz, K. Ashraf, S. Han, W.J. Dally, K. Keutzer, Squeezenet: alexnet-level accuracy with 50x fewer parameters and <1mb model size. CoRR **abs/1602.07360** (2016), http://arxiv.org/abs/1602.07360
20. H. Ismail Fawaz, G. Forestier, J. Weber, L. Idoumghar, P.A. Muller, Deep learning for time series classification: a review. Data Mining Knowl. Discov. **33**(4), 917–963 (2019)
21. H. Jin, Q. Song, X. Hu, Auto-keras: an efficient neural architecture search system, in *Proceedings of the 25th ACM SIGKDD International Conference on Knowledge Discovery and Data Mining*, pp. 1946–1956 (2019)
22. A. Karpathy, L. Fei-Fei, Deep visual-semantic alignments for generating image descriptions. IEEE Trans. Pattern Analysis Mach. Intell. **39**(4), 664–676 (2017), https://doi.org/10.1109/TPAMI.2016.2598339
23. R. Korichi, M. Guillemot, C. Heusèle, Tuning neural network hyperparameters through bayesian optimization and application to cosmetic formulation data, in *ORASIS 2019* (2019)
24. A. Krizhevsky, I. Sutskever, G.E. Hinton, Imagenet classification with deep convolutional neural networks, in *26th Annual Conference on Neural Information Processing Systems 2012*, eds by P.L. Bartlett, F.C.N. Pereira, C.J.C. Burges, L. Bottou, K.Q. Weinberger, pp. 1106–1114. Lake Tahoe, Nevada, United States, 3–6 Dec 2012. Advances in Neural Information Processing Systems 25. http://papers.nips.cc/paper/4824-imagenet-classification-with-deep-convolutional-neural-networks
25. G. Larsson, M. Maire, G. Shakhnarovich, Fractalnet: ultra-deep neural networks without residuals, in *International Conference on Learning Representations*. OpenReview.net (2017) https://openreview.net/forum?id=S1VaB4cex
26. Y. LeCun, L. Bottou, Y. Bengio, P. Haffner, Gradient-based learning applied to document recognition. Proc. IEEE **86**(11), 2278–2324 (1998)
27. Y. LeCun et al., Generalization and network design strategies. Connect. Perspect. **19**, 143–155 (1989)
28. M. Lin, Q. Chen, S. Yan, Network in network, in *2nd International Conference on Learning Representations, ICLR 2014* eds by Y. Bengio, Y. LeCun (eds.) , Banff, AB, Canada, 14–16 April 2014. Conference Track Proceedings (2014), http://arxiv.org/abs/1312.4400
29. H. Liu, K. Simonyan, Y. Yang, Darts: differentiable architecture search. arXiv preprint arXiv:1806.09055 (2018)
30. J. Long, E. Shelhamer, T. Darrell, Fully convolutional networks for semantic segmentation, in *2015 IEEE Conference on Computer Vision and Pattern Recognition, CVPR 2015*, pp. 3431–3440. Boston, MA, USA, 7–12 June 2015. IEEE Computer Society (2015). https://doi.org/10.1109/CVPR.2015.7298965
31. W.S. McCulloch, W. Pitts, A logical calculus of the ideas immanent in nervous activity. Bull. Math. Biophys. **5**(4), 115–133 (1943)

32. H. Noh, S. Hong, B. Han, Learning deconvolution network for semantic segmentation, in *2015 IEEE International Conference on Computer Vision, ICCV 2015*, pp. 1520–1528. Santiago, Chile, 7–13 December 2015. IEEE Computer Society (2015). https://doi.org/10.1109/ICCV.2015.178
33. R. Pascanu, C. Gülçehre, K. Cho, Y. Bengio, How to construct deep recurrent neural networks, in *2nd International Conference on Learning Representations, ICLR 2014*, eds by Y. Bengio, Y. LeCun. Banff, AB, Canada, 14–16 April 2014. Conference Track Proceedings (2014). http://arxiv.org/abs/1312.6026
34. H. Pham, M.Y. Guan, B. Zoph, Q.V. Le, J. Dean, Efficient neural architecture search via parameter sharing. arXiv preprint arXiv:1802.03268 (2018)
35. J.B. Pollack, Recursive distributed representations. Artif. Intell. **46**(1–2), 77–105 (1990)
36. W. Rawat, Z. Wang, Deep convolutional neural networks for image classification: a comprehensive review. Neural Comput. **29**, 1–98 (2017)
37. E. Real, A. Aggarwal, Y. Huang, Q.V. Le, Aging evolution for image classifier architecture search, in *Proceedings of the Thirty-Third AAAI Conference on Artificial Intelligence (AAAI-19)*. (AAAI Press/IJCAI, 2019)
38. E. Real, A. Aggarwal, Y. Huang, Q.V. Le, Regularized evolution for image classifier architecture search, in *Proceedings of the Thirty-Third AAAI Conference on Artificial Intelligence (AAAI-19)*. vol. 33, pp. 4780–4789. AAAI Press/IJCAI (2019)
39. F. Rosenblatt, The perceptron: a probabilistic model for information storage and organization in the brain. Psychol. Rev. **65**(6), 386 (1958)
40. D.E. Rumelhart, G.E. Hinton, R.J. Williams, Learning representations by back-propagating errors. Nature **323**(6088), 533–536 (1986)
41. E. Shelhamer, J. Long, T. Darrell, Fully convolutional networks for semantic segmentation. IEEE Trans. Pattern Analysis Mach. Intell. **39**(4), 640–651 (2017)
42. K. Simonyan, A. Zisserman, Very deep convolutional networks for large-scale image recognition, in *3rd International Conference on Learning Representations, ICLR 2015*, eds by Y. Bengio, Y. LeCun. San Diego, CA, USA, 7–9 May 2015, Conference Track Proceedings (2015). http://arxiv.org/abs/1409.1556
43. J. Snoek, H. Larochelle, R.P. Adams, Practical bayesian optimization of machine learning algorithms, in *26th Annual Conference on Neural Information Processing Systems 2012* eds by P.L. Bartlett, F.C.N. Pereira, C.J.C. Burges, L. Bottou, K.Q. Weinberger, pp. 2960–2968. Advances in Neural Information Processing Systems 25. Proceedings of a meeting held 3–6 December 2012, Lake Tahoe, Nevada, United States (2012). http://papers.nips.cc/paper/4522-practical-bayesian-optimization-of-machine-learning-algorithms
44. R. Socher, E.H. Huang, J. Pennington, A.Y. Ng, C.D. Manning, Dynamic pooling and unfolding recursive autoencoders for paraphrase detection, in *25th Annual Conference on Neural Information Processing Systems 2011*, eds by J. Shawe-Taylor, R.S. Zemel, P.L. Bartlett, F.C.N. Pereira, K.Q. Weinberger, pp. 801–809. Advances in Neural Information Processing Systems 24. Proceedings of a meeting held 12–14 December 2011, Granada, Spain (2011). http://papers.nips.cc/paper/4204-dynamic-pooling-and-unfolding-recursive-autoencoders-for-paraphrase-detection
45. R. Socher, A. Perelygin, J. Wu, J. Chuang, C.D. Manning, A.Y. Ng, C. Potts, Recursive deep models for semantic compositionality over a sentiment treebank, in *Proceedings of the 2013 Conference on Empirical Methods in Natural Language Processing, EMNLP 2013*, pp. 1631–1642. Grand Hyatt Seattle, Seattle, Washington, USA, 18–21 October 2013. A meeting of SIGDAT, a Special Interest Group of the ACL. ACL (2013). https://www.aclweb.org/anthology/D13-1170/
46. C. Szegedy, W. Liu, Y. Jia, P. Sermanet, S.E. Reed, D. Anguelov, D. Erhan, V. Vanhoucke, A. Rabinovich, Going deeper with convolutions, in *2015 IEEE Conference on Computer Vision and Pattern Recognition, CVPR 2015*, pp. 1–9. Boston, MA, USA, 7–12 June 2015. IEEE Computer Society (2015). https://doi.org/10.1109/CVPR.2015.7298594

47. A. Toshev, C. Szegedy, Deeppose: human pose estimation via deep neural networks, in *2014 IEEE Conference on Computer Vision and Pattern Recognition, CVPR 2014*, . pp. 1653–1660. Columbus, OH, USA, 23–28 June 2014. IEEE Computer Society (2014). https://doi.org/10.1109/CVPR.2014.214
48. B. Widrow, M.E. Hoff, *Adaptive Switching Circuits* (Stanford Univ Ca Stanford Electronics Labs, Tech. rep., 1960)
49. S. Xie, R.B. Girshick, P. Dollár, P., Tu, Z., He, K.: Aggregated residual transformations for deep neural networks, in *2017 IEEE Conference on Computer Vision and Pattern Recognition, CVPR 2017*, pp. 5987–5995. Honolulu, HI, USA, 21–26 July 2017. IEEE Computer Society (2017). https://doi.org/10.1109/CVPR.2017.634
50. X.S. Yang, Chapter 6—genetic algorithms, in *Nature-Inspired Optimization Algorithms*, 2nd edn, eds by X.S. Yang, pp. 91–100. (Academic Press, 2021). https://doi.org/10.1016/B978-0-12-821986-7.00013-5, https://www.sciencedirect.com/science/article/pii/B9780128219867000135
51. T. Yu, H. Zhu, Hyper-parameter optimization: a review of algorithms and applications. arXiv preprint arXiv:2003.05689 (2020)
52. B. Zoph, V. Vasudevan, J. Shlens, Q.V. Le, Learning transferable architectures for scalable image recognition, in *2018 IEEE Conference on Computer Vision and Pattern Recognition, CVPR 2018*, pp. 8697–8710. Salt Lake City, UT, USA, 18–22 June 2018. IEEE Computer Society (2018). https://doi.org/10.1109/CVPR.2018.00907

Accelerating Deep Neural Networks with Phase-Change Memory Devices

Katie Spoon, Stefano Ambrogio, Pritish Narayanan, Hsinyu Tsai, Charles Mackin, An Chen, Andrea Fasoli, Alexander Friz, and Geoffrey W. Burr

Abstract In this chapter, we discuss recent advances in the hardware acceleration of deep neural networks with analog memory devices. Analog memory offers enormous potential to speed up computation in deep learning. We study the use of Phase-Change Memory (PCM) as the resistive element in a crossbar array that allows the multiply-accumulate operation in deep neural networks to be performed in-memory. With this promise comes several challenges, including the impact of conductance drift on deep neural network accuracy. Here we introduce popular neural network architectures and explain how to accelerate inference using PCM arrays. We present a technique to compensate for conductance drift ("slope correction") to allow in-memory computing with PCM during inference to reach software-equivalent deep learning baselines for a broad variety of important neural network workloads.

1 Introduction

Today's world has a high demand for an ability to quickly make sense of a rapidly expanding flow of data [1]. In this context, machine learning techniques have become widely popular to help extract meaningful information from a variety of data such as images, text and speech [2]. In the last decade, the confluence of large amounts of labelled datasets, reliable algorithms such as stochastic gradient descent and increased computational power from CPUs (Central Processing Units) and GPUs (Graphics Processing Units) has enabled deep learning, a branch of machine learning, to revolutionize many fields, from image classification to speech recognition to language translation [3–6].

Neural networks typically have input and output layers, with many hidden layers in between. Each layer contains many neurons, where the output of each neuron

K. Spoon · S. Ambrogio (✉) · P. Narayanan · H. Tsai · C. Mackin · A. Chen · A. Fasoli · A. Friz · G. W. Burr
IBM Research-Almaden, 650 Harry Road, San Jose, CA 95120, USA
e-mail: stefano.ambrogio@ibm.com

K. Spoon
e-mail: katherine.spoon@colorado.edu

R. Micheloni and C. Zambelli (eds.), *Machine Learning and Non-volatile Memories*,
https://doi.org/10.1007/978-3-031-03841-9_3

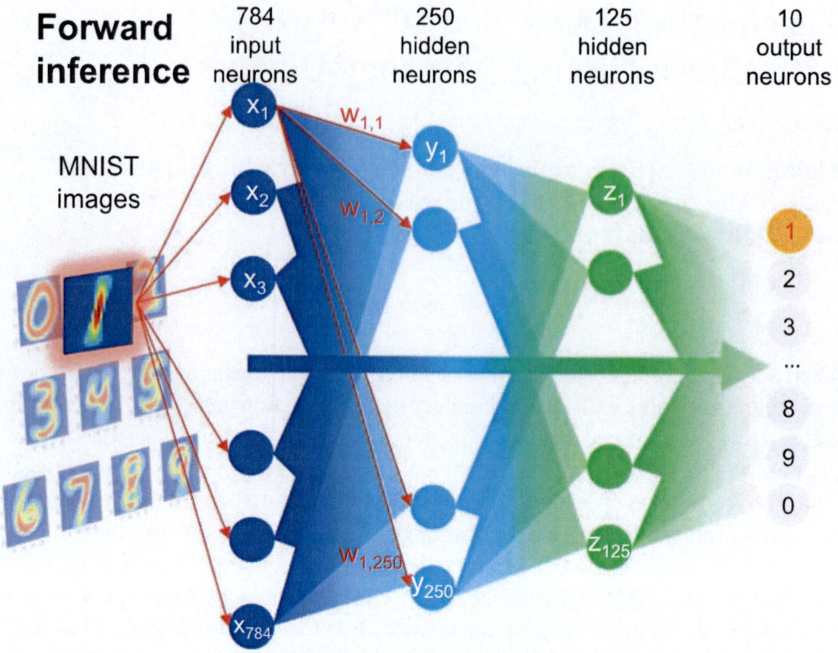

Fig. 1 A simple task is to classify images of handwritten digits from the MNIST dataset, where the network's goal is to output the number the image contains. The input layer has 784 input neurons, the number of pixels in a 28 × 28 image. In the network shown, there are two hidden layers with 250 and 125 neurons, and a final layer with 10 output neurons, one for each possible classification (digits 0 through 9). There are connections between every pair of neurons in one layer to the next: $w_{1,1}$ is the weight from x_1 to y_1, $w_{1,2}$ is the weight from from x_1 to y_2, and so on. During forward inference, given an image, e.g. an image of a "one", the example proceeds through the trained network, which should select the output "1" out of the 10 choices. Adapted with permission from [7]. Copyright 2017 IEEE

is determined by the weights feeding into that neuron, passed through a non-linear activation function [1]. The simplest neural network is a fully connected network, where every neuron in one layer is connected to every neuron in the next layer, as shown in Fig. 1. This is also referred to as a Multi-Layer Perceptron (MLP). In this particular example, a very simple dataset is used: the MNIST (Modified National Institute of Standards and Technology) database of handwritten digits [6]. Here, the goal is to train the network to recognize input digits using what is called a training dataset, followed by testing the network's ability to classify previously unseen images from the test dataset [2].

While MNIST classification provides limited challenges, more complicated network architectures and datasets have recently been used in computer vision to solve more interesting and challenging tasks. One of the best-known is the ImageNet problem, where the goal is to classify real-world images into one thousand categories (cat, dog, etc.) based on their content [8]. By 2015, the best deep neural network was

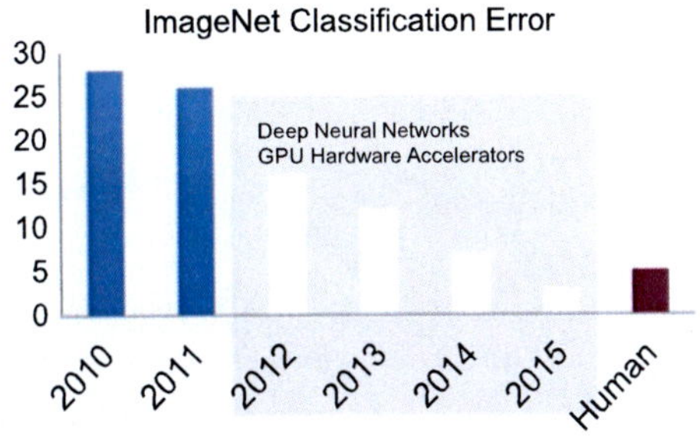

Fig. 2 Neural networks existed long before 2015, however, the explosive switch to deep learning occurred due to the compute power and large data sets that became more widely available in recent years

making so few classification errors that it had effectively surpassed human ability, as shown in Fig. 2. This was largely due to the accelerated training of a large convolutional neural network using high performance GPUs [9]. The main advantage of GPUs resides in the highly parallel and efficient vector-matrix multiplication, which constitutes the core of neural network computations. This improvement has recently enabled the training of increasingly larger networks with millions or even billions of adjustable weights, providing increasingly higher accuracy classification and recognition performances [3, 10].

1.1 The Promise of Analog AI

Deep neural networks are attractive to hardware designers due to the nature of the core operation in a neural network: the vector-matrix product. Today this operation is typically performed by digital accelerators (i.e. GPUs), which are set up as shown in Fig. 3a, with the processor on one side and memory on the other, connected by a bus. This is known as the Von Neumann Architecture [11]. The bus can become a bottleneck since the data is sent back and forth from memory to processor, therefore a limited amount of data can be moved at any time in the communication bus [3, 10, 12]. Analog accelerators, on the other hand, perform computations directly in memory (Fig. 3b), which also behaves as a processor [13–19]. In the context of neural networks, analog accelerators offer a more natural implementation of fully connected layers. Since non-volatile memory (NVM) devices are organized in crossbar arrays, which provide a full connection between every input and every output, the mapping of fully connected layers into memory arrays becomes straightforward, thus increasing the density of programmable weights, the computation efficiency and the processing speed [13, 20].

How is the network mapped into memory? There are many different choices for the resistive element of the crossbar array, and in this work we use Phase Change Memory (PCM) [21, 22]. The fully connected neural network in Fig. 1 can be mapped onto crossbars arrays with the neurons stored in peripheral circuitry and the weights

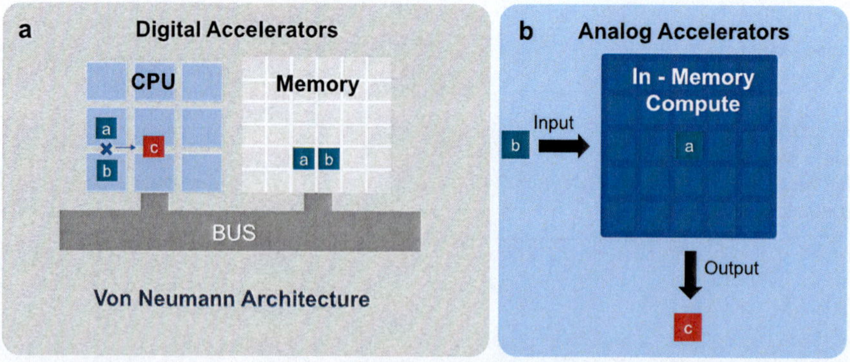

Fig. 3 Suppose we need to multiply two numbers, **a** and **b**. **a** In a typical setup, **a** and **b** start sitting in memory. They are both sent to the processor, and the answer **c** is computed and sent back to the memory. This consumes a lot of energy in data movement. Additionally, the bus can become a bottleneck. **b** Analog accelerators perform the computation in memory by storing **a** in the memory (crossbar array) and sending **b** in as a voltage, producing the answer **c** in-place

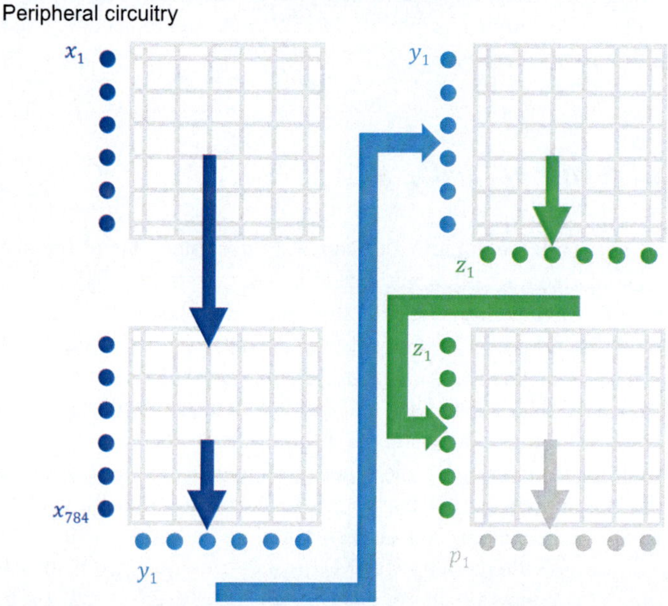

Fig. 4 The simple MLP from Fig. 1 has an input layer, two hidden layers, and an output layer. These can be mapped onto crossbar arrays of size 512 × 512, for example. Since the input layer is 784 × 250, two arrays are needed for the first layer, with 784 input neurons representing the rows and 250 hidden neurons representing the columns. There is an additional crossbar array for the 250 × 125 hidden layer, and a final array for the 125 × 10 output layer. The arrows represent the signal propagation through the network during forward inference

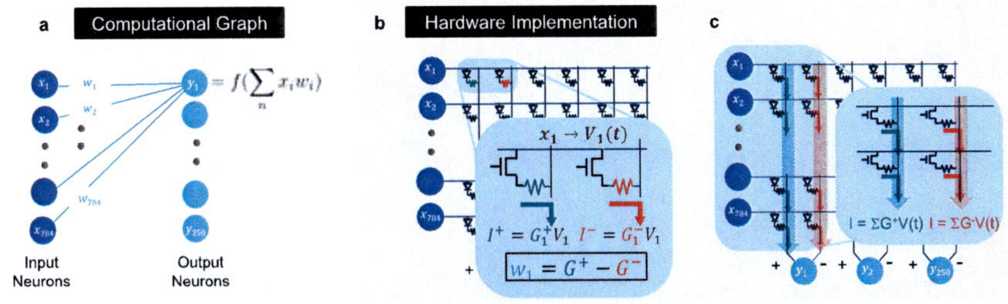

Fig. 5 **a** To compute the output of a neuron, for example, y_1, the inputs x times the weights w are summed, then an activation function f is applied. This function could be the sigmoid mentioned in the discussion of the LSTM network in Sect. 1.2, or any other number of functions. **b** Ohm's law is used to multiply $x * w$, where x is represented as a voltage $V(t)$, and (c) Kirchoff's law is used to accumulate (sum) the results along the columns

of the neural network stored in the crossbar array (Fig. 4). During forward inference, the signal propagates through the crossbar arrays all the way to the output [16].

Figure 5 shows how a single crossbar array (for example, the first layer) is implemented in analog hardware. The essential calculation is the **multiply-accumulate operation** as shown in Fig. 5a. To perform this operation in hardware, the weights of the network are encoded as the difference between a pair of conductances G+ and G- [13], which are analog tuned using proper programming schemes [20, 23]. Values of the input neuron x are encoded as voltages $V(t)$ and applied to the elements in a row. The multiplication operation between the x and w terms is performed through Ohm's law (Fig. 5b). There are two possible ways to implement the input x, either by keeping the pulse-width constant and tuning the voltage amplitude, or by encoding the x value in the pulse duration and keeping the voltage amplitude constant. While the first scheme allows better time control since all pulses show identical duration, the second scheme helps counteract the read voltage non-linearity that all NVM devices experience, and prevents unwanted device programming for excessive read amplitudes. By Kirchoff's current law, all of the product terms generated on a single column by all devices are accumulated in terms of aggregate current (Fig. 5c) to produce the final result at the bottom of the array in the form of an accumulated charge on a peripheral capacitor. Through this process, the multiply-accumulate operation, which is the most computationally expensive operation performed in a neural network, can be done in constant time, without any dependence on neural network layer size [13].

If these multiply-accumulate operations needed to be done with 32-bit or even 16-bit floating-point precision, this approach would probably not be feasible due to intrinsic noisy operations such as PCM read or write, exact weight programming or peripheral circuit conversion precision. But research on deep neural networks has shown that we can reduce the precision of weights and activations down to 8-bit integers, or even lower, without significant consequence to network accuracy [24]. For this reason, analog AI is a promising approach for these multiply-accumulate

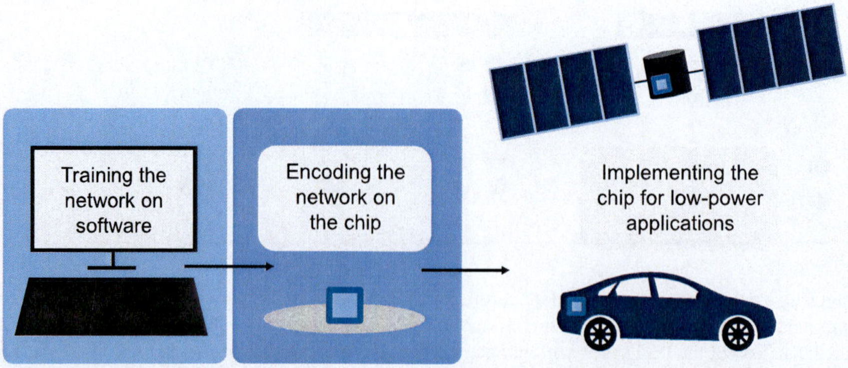

Fig. 6 The goal is to train a neural network in software, then encode those trained weights of the network onto a chip, then implement the chip for low-power applications, such as in a self driving car

computations in the context of deep neural networks, due to the intrinsic resilience that such networks provide to noisy operations [14].

There are two main stages of any deep learning pipeline: training and inference. During training, the network's weights are updated, via back-propagation [2, 6], to better distinguish between different labeled data examples. Once the model is trained, it can be used to predict the labels of new examples during forward inference. There are hardware opportunities for both training and inference, but in this chapter we focus on the design of forward inference chips that can run quickly at low power (Fig. 6).

1.2 Two Common Neural Network Architectures

There are many applications of deep learning and, among those, computer vision and natural language processing are two of the most notable. It is helpful to use different types of neural networks depending on the structure of the input data, and, as an example, images and text have very different structures. Additionally, the MNIST handwritten digit task in Fig. 1 is very simple, and the most interesting problems today are difficult, requiring larger and more sophisticated networks. In this section we briefly review two of the most popular neural networks used in each of these two areas [25].

Computer Vision: Convolutional Neural Networks (CNNs) A slightly more challenging task than classifying handwritten digits is image classification into categories, like birds and dogs. A convolutional neural network (CNN) differs from the fully connected network in Fig. 1 by using convolutions to better process the image data. The network utilizes a set of filters that scan across the image, taking in a small amount of information at a time. The CNN shown in Fig. 7 has one convolutional

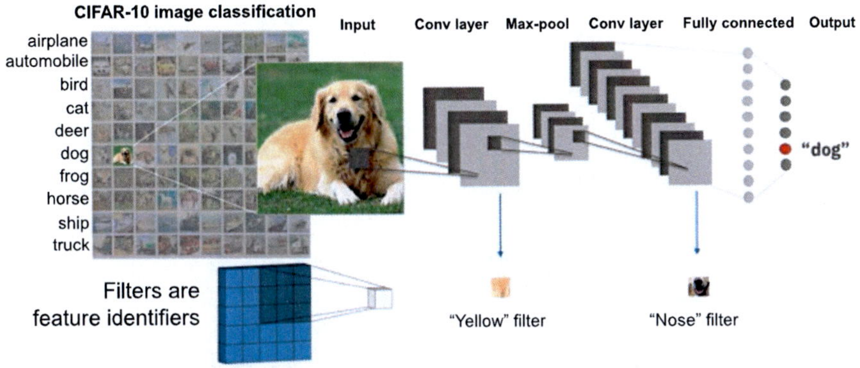

Fig. 7 In this example, the goal is to classify an image of a dog correctly out of the ten classes of images in the CIFAR10 dataset [26]. The input to the network is a tensor of 3 dimensions: [image-width, image-height, color-depth]. For instance, a single color image can be considered as three closely-related 2D images, one each for the red, green, and blue channels. The output from the network is a 1D vector, with one element for each class (dog, cat, horse, and so on)

layer followed by a max-pooling layer that reduces the dimensionality and determines the most important information from the previous layer, followed by another convolutional layer and a fully connected layer at the end to classify the image.

Each filter can be thought of as a feature identifier. For example, the earlier features might be the lines and colors of the dog, whereas later filters may represent very specific features, like noses or ears. It is important to note that these filters are **not designed by humans**; in a deep neural network, these filters are learned, as the filter weights are slowly updated during repeated exposure to the very large training dataset. These filters allow CNNs to successfully capture the spatial and temporal dependencies in the image, loosely mimicking the human vision system.

In reality, much deeper CNNs are used for challenging tasks. For example, ResNet [27] still has filter sizes of 3×3, however instead of only 4 filters in the first layer, there are 64 filters for the first layer, 64 for the next, and so on (see Fig. 8).

Natural Language Processing: Recurrent Neural Networks (RNNs) To recognize text, different strategies are adopted, given the sequential nature of the input data. For example, assume we want to build a chatbot, that, given a query, will route the user to the correct category of answers. If a user asks "How do I request an account?", the chatbot should classify the question as a "new account" question. This is a very different application from image classification. CNNs *could* be used for this problem, but it makes more sense to use a network that took advantage of the **sequential nature of text** as the input.

First, the sentence is split into words (tokens) to be sent through the network one by one (Fig. 9a). The first word is the first input to the hidden layer (Fig. 9b). This produces an output, similar to the fully connected network in Fig. 1. The second word is where the recurrence comes in. When "do" is sent into the network, the output is calculated using both the current hidden state as well as the output of the

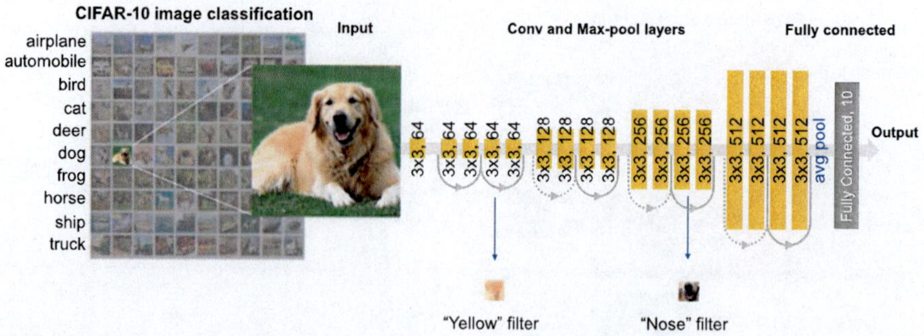

Fig. 8 ResNet is a much deeper network than the standard CNN in Fig. 7, so information can sometimes be lost in deeper layers. The arrows show residual connections, designed to help the information stay relevant deeper in the network (hence the "Res" in "ResNet"). One of the 64 early filters could be a "yellow" identifier, and one of the later filters could be a "ear" identifier, for example

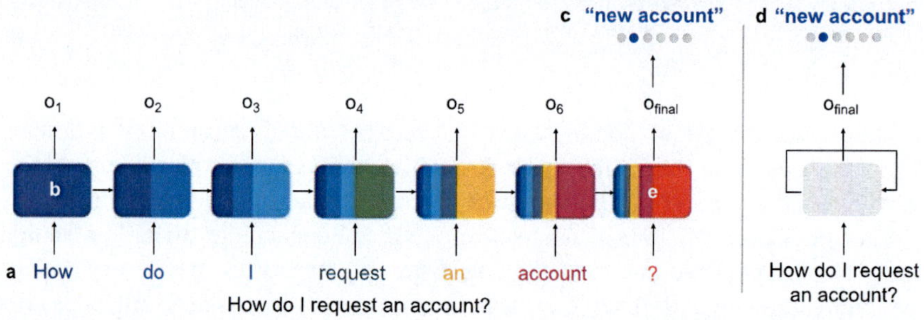

Fig. 9 An unrolled recurrent neural network (RNN) processing a single input sequence query is shown. **a** The sequence is split into words or tokens and they are sent into the network sequentially, starting by **b** sending the first word "How" through the hidden state and producing an output o_1. On the next word, "do", the previous hidden state and the current input produce the next output. This continues until the entire sequence has been processed, then **c** the final output though the fully connected layer produces the output class. The network has determined that this query, "How do I request an account?" is a "new account" question. **d** Currently the network is unrolled, but it can be represented rolled up, with an arrow representing the hidden state feeding back into itself before finishing the sequence. **e** Recurrent neural networks are useful for processing sequential data, but they can lose information over time. For example, in the last hidden state, information from the first hidden state is only represented by a small sliver

previous hidden state. This continues throughout inference of the sequence, until the final hidden state output is fed to a fully connected layer that classifies the query as a "new account" question (Fig. 9c).

Unfortunately, RNNs lose information in long sequences - they have "short-term memory". The last hidden state illustrates this: it contains only a small sliver of the earlier words of the sentence, largely depending instead on the last word or symbol, in this case the question mark (Fig. 9e).

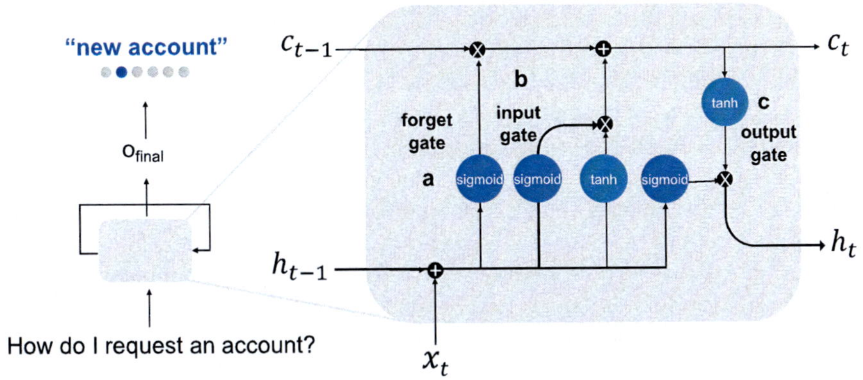

Fig. 10 An LSTM network is still a recurrent neural network (RNN), with the only difference being the structure of the hidden state. Here a closeup of the cell shows the various gates the input goes through before exiting the hidden state. For example, the first input x is "How", for time step $t = 0$. The next word, "do" is $x_{t=1}$. There are three gates: **a the forget gate**, determining which information from the previous hidden state should be forgotten based on the current input, **b the input gate**, deciding which values to update in the cell state (maintained along the top), and finally **c the output gate**, producing the final hidden state value

To solve this problem, a modified RNN called a **Long Short-Term Memory Network (LSTM)** is often used. The hidden state is replaced by a more sophisticated cell that still unrolls for each new word, however, it also contains a sequence of gates that determine which information from the new input token and the old cell state are relevant and which should be thrown out. This adjustment has allowed for major advancements in natural language processing applications.

In a closeup of this cell in Fig. 10, instead of weighting both the current input and the old hidden state equally, three gates determine how much to keep or throw away.

Information is maintained along the top, in what is called the cell state, or c_t. The different gates determine how to update the cell state. The first gate is the forget gate (Fig. 10a), which uses a sigmoid function to process the input and previous hidden state and determine which information in the cell state should stay and what should be forgotten. For example, it could be that during training, the network has determined that the word "account" is really important for prediction, in which case the forget gate would output a 1, indicating to fully keep that information.

In general, the sigmoid function outputs numbers between zero and one, identifying which elements should be let through. A value of zero means "keep nothing", whereas one means "keep everything". A tanh function outputs numbers between -1 and 1 and is used to help regulate the values flowing through the network.

The input gate (Fig. 10b) determines which values to add to the cell state, and a tanh function applied to the same input generates the new candidate values that could be added based on the current input. To update the cell state, first the previous cell state c_{t-1} is multiplied by the forget gate output. Then the result is added to the input gate times the candidate values provided by the tanh function. Last, the output gate (Fig. 10c) determines the output of the cell.

Unlike the multi-dimensional tensors in the image application, the cell-state, hidden-state, inputs and outputs are all one-dimensional vectors, and the multiplies and adds within the LSTM are performed in an element-wise fashion. As a result, a natural language processing system will often include an encoder—to turn words in English or another language into a vector of floating-point numbers—and a decoder, to convert each output vector into a set of predicted probabilities identifying what the most probable next word in the sentence is. By choosing the right encoder and decoder with careful training, an LSTM can be used for machine translation from one language to another.

2 Software-Equivalent Accuracy in Analog AI

While the implementation of CNNs and LSTMs using crossbar arrays of PCM devices can potentially provide fast and energy efficient operations, Phase Change Memory also presents many non-idealities that need to be corrected in order to reach software-equivalent accuracy with analog hardware.

2.1 Programming Strategies

To accurately perform neural network inference, weights must be precisely programmed by tuning the conductance of the devices. Write and read noise affect all NVM types, while PCM also experiences an additional non-ideality, **conductance drift**, namely a reduction of conductance over longer times, which degrades the computation precision, eventually decreasing the classification accuracy [28, 29]. Even small changes from the trained weights to the weights programmed on the chip can have a crucial impact on accuracy. Ideally, the relationship between programming pulse and achieved conductance state should be predictable, where a certain number and shape of pulses will always program a certain conductance value G. However, actual programming traces show a large variability, as can be seen from simulated traces in Fig. 11, with each device behaving slightly differently from the others (inter-device variability). Even a single device experiences different conductance traces under the same programming conditions (intra-device variability), potentially causing a drop in neural network accuracy [30].

In order to make the programming process more precise, a closed-loop tuning (CLT) procedure can be adopted, and a more complex unit cell is often used that contains not only one most significant pair (MSP) of conductances G+ and G-, but also includes a least significant pair g+ and g- (see Fig. 12) [14, 23, 30]. This CLT operation consists of tuning the weights in four phases, one for each of the four conductances G+, G-, g+ and g-. To accelerate the write speed, programming can be

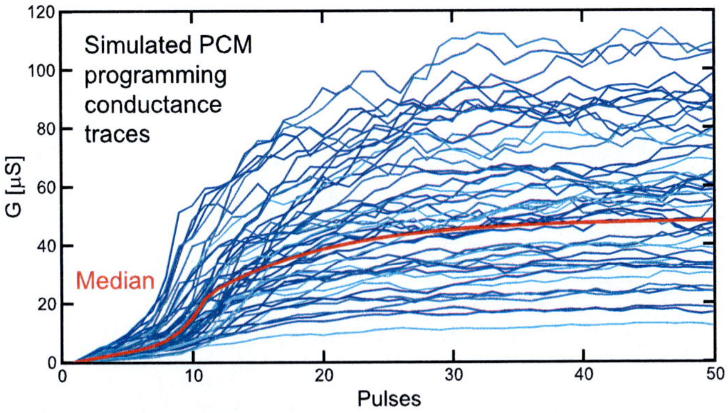

Fig. 11 Each simulated curve represents a different device programming behavior. Ideally we want to be able to predict and better control these trajectories. Adapted with permission from [23]. Copyright 2019 John Wiley and Sons

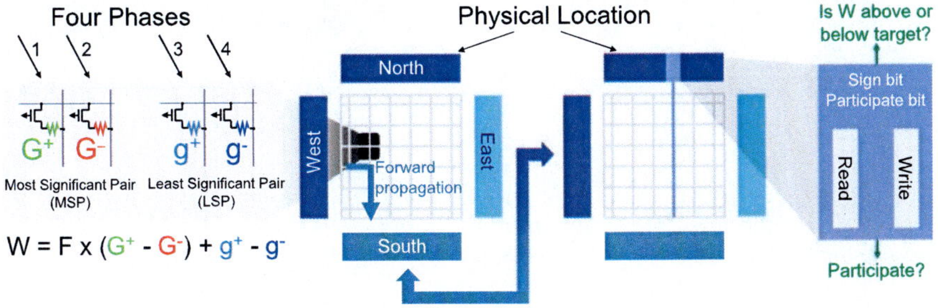

Fig. 12 The unit cell contains four conductance values, two for the most significant pair (MSP): G+ and G-, and two for the least significant pair (LSP): g+ and g-. The programming strategy focuses on each one of these values in turn, iteratively updating the weight value depending on how close it is to the target weight (the goal weight). Each column additionally contains two pieces of information: whether or not the current weight is above or below the target weight, along with whether or not this column will participate in the current programming phase. Adapted with permission from [23]. Copyright 2019 John Wiley and Sons

performed in a row-wise parallel fashion, tuning all weights in one row at the same time. To enable this, each weight is checked against the target after each read step. If a weight reaches, or overcomes, the target weight, further programming needs to be prevented. Each column contains a sign bit and a participation bit. The sign bit represents whether or not the full weight W is above or below the target weight in the current stage, and the participation bit represents whether that column should participate in the current phase or stop [23].

The error is defined as the difference between the actual programmed weight and the target weight. The error should go to 0 throughout the four phases. For positive (negative) target weights, Phase 1 starts by applying pulses to G+ (G-) until the weight exceeds the threshold. Then the process is repeated for G- (G+), this time applying

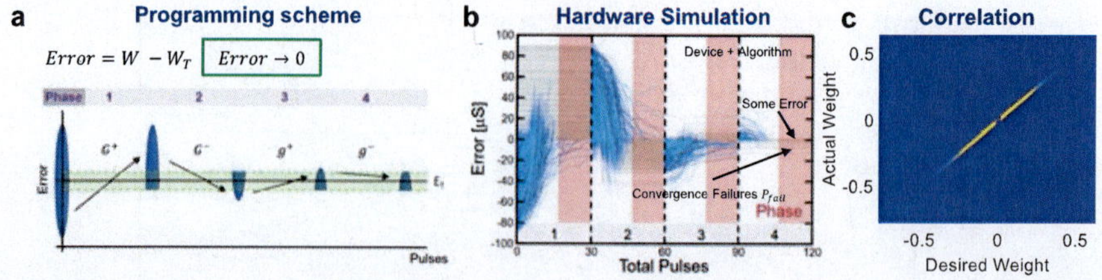

Fig. 13 **a** Iterative programming strategy that successively tunes the four conductance in a weight, **b** leading to **c** a strong correlation between the desired programmed weights and the actual programmed weights. Adapted with permission from[23]. Copyright 2019 John Wiley and Sons

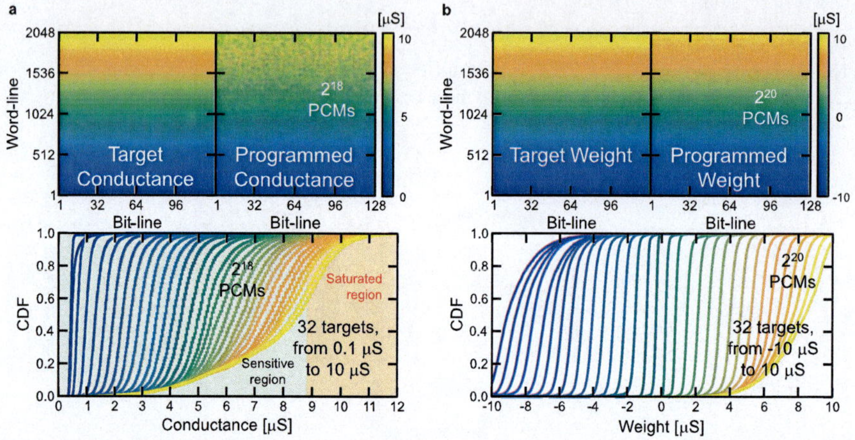

Fig. 14 Experimental comparison between single PCM (**a**) and full weight (**b**) closed-loop tuning. Top figures show the target conductances and weights, together with the achieved programmed results. Bottom figures show the corresponding cumulative distribution functions (CDF) for variable targets. While single-device CDFs (**a**) reveal increasingly broadened distributions due to programming error and device saturation for increasingly larger targets, weight CDFs (**b**) reveal steeper curves, which is a consequence of the improved programming precision, and reduced saturation due to increased weight dynamic range. Adapted with permission from [29]. Copyright 2019 IEEE

pulses until it drops below the threshold. This successive approximation technique continues for the LSP (Fig. 13a). In this case, since the contribution from g+ and g- is reduced by a constant factor F around 2–4, the program precision increases. By simulating this process, Fig. 13b shows that ∼98% of weights can be programmed effectively using this strategy, with only ∼2% of weights out of target, thus providing a very strong correlation between the programmed and target weights (Fig. 13c).

To verify the impact of MSP and LSP pairs on write precision, actual experiments are shown in Fig. 14. The lower conductance values are easier to reach, but as the target value gets higher, the PCMs less reliably program the desired value (Fig. 14a) due to variability in the maximum achievable conductance for each device [29]. This can also be seen by studying the cumulative distribution functions (CDF), which show that the percentage of PCM devices that reach a certain conductance target decreases for increasing target values. In addition, CDF curves are not very steep due to programming errors that limit the precision we can achieve in tuning a single PCM device.

When the weights are split into four conductances with a simplified version of the programming strategy described, more of the PCMs reach the desired conductance values (Fig. 14b) due to an increased dynamic range. In addition, CDF distributions are steeper, revealing a better control of the weight programming, leveraging both MSP and LSP conductance pairs [29]. Using this programming strategy, software-equivalent accuracy was demonstrated, with a mixed software-hardware experiment, on LSTM networks similar to the one introduced in Sect. 1.2 [29].

2.2 *Counteracting Conductance Drift Using Slope Correction*

Additionally, even after programming the desired values, phase change memory exhibits another non-ideality called conductance drift, where the device conductance decays with increasing time due to the structural relaxation of the amorphous phase [28]. Since weights are encoded in the conductance state of PCM devices, drift degrades the precision of the encoded weight, and needs to be counteracted [31, 32].

Drift typically affects reset and partial reset states, with an empirical power law describing the time dependence:

$$G(t) = G_0 * \left(\frac{t}{t_0}\right)^{-\nu},$$

where G_0 is the very first conductance measurement obtained at time t_0 after the programming time. For increasing times t, conductance decays with a power law defined by a negative drift coefficient ν, indicating the drift rate.

Drift is a very rapid process right after programming but slows down considerably as time continues. While all PCM devices experience drift, each device drifts at a slightly different rate. To properly evaluate the drift coefficient distribution across multiple PCMs, ν coefficients for 20,000 devices were extracted by measuring the conductances at multiple times over 32 h [31]. Figure 15a shows the ν distribution for all 20,000 devices as a function of the initial conductance measurement $G(t_1)$. The corresponding median $\bar{\nu}$ and standard deviation σ_ν are then extracted and shown in Fig. 15b. As a first approximation, we can consider $\bar{\nu}$ equal to 0.031, and σ_ν equal to 0.007. These parameters have been implemented in our model to study the impact of drift on neural network inference [31].

In order to understand how to correct conductance drift, it can be helpful to look more carefully at the activation function used during the multiply-accumulate operation discussed earlier. This activation function f (also known as a squashing function) transforms the data. Different functions will result in different transforms.

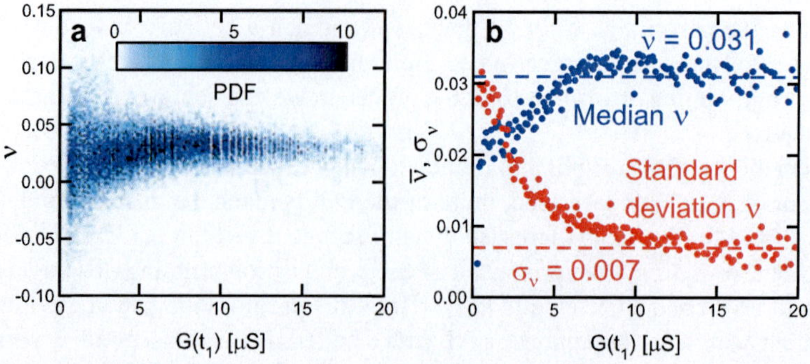

Fig. 15 Experimental drift characterization on 20,000 PCM devices. After PCM programming, conductances have been measured up to 32 h, and corresponding ν have been extracted and plotted as a function of the initial conductance (**a**). The extracted median $\bar{\nu}$ and standard deviation σ_ν are plotted in (**b**). Adapted with permission from [31]. Copyright 2019 IEEE

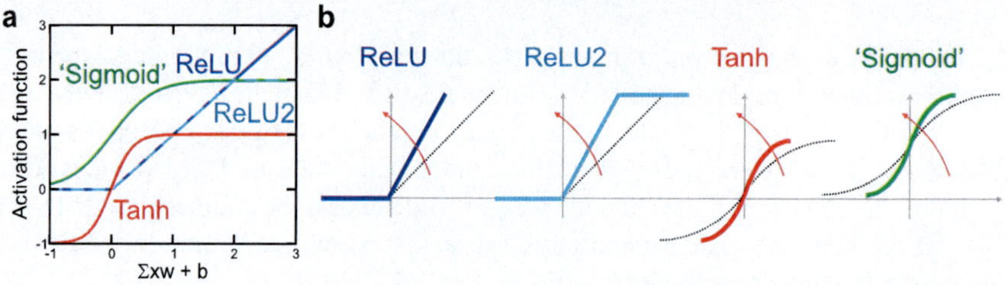

Fig. 16 A wide variety of activation functions are used in practice, four examples of which are shown in (**a**). The sigmoid function is expanded to better compare to the tanh function. **b** We can apply activation amplification factors ("slope correction" factors) to these functions to counteract drift over time. Adapted with permission from [31]. Copyright 2019 IEEE

Here four main functions are introduced: ReLU (Rectified Linear Unit), clamped ReLU, rescaled sigmoid, and tanh, as shown in Fig. 16a. Except for ReLU, which is unbounded, all the other squashing functions are studied here at the same amplitude. As the standard deviation of the drift coefficients approaches 0, the effects of drift can be factored out of the multiply-accumulate operation as a single time-varying constant. This allows for the compensation of conductance drift by applying a time-dependent activation amplification equal to $(t/t_0)^{+\nu_{correction}}$, where the previously extracted $\bar{\nu}$ is chosen as $\nu_{correction}$. This factor is applied to the slope of the activation function (Fig. 16b) [31].

This technique has been evaluated on several networks. Initially, without any correction, the fully connected network experiences a marked accuracy degradation, over time, on the MNIST dataset of handwritten digits (Fig. 17a), due to conductance drift. With slope correction, however, results for all the activation functions studied improve significantly (Fig. 17b). The technique has also been evaluated on a ResNet-18 CNN trained on the CIFAR-10 image classification dataset (Fig. 18a) and an LSTM, trained on the text from the book Alice in Wonderland (Fig. 18b). With these

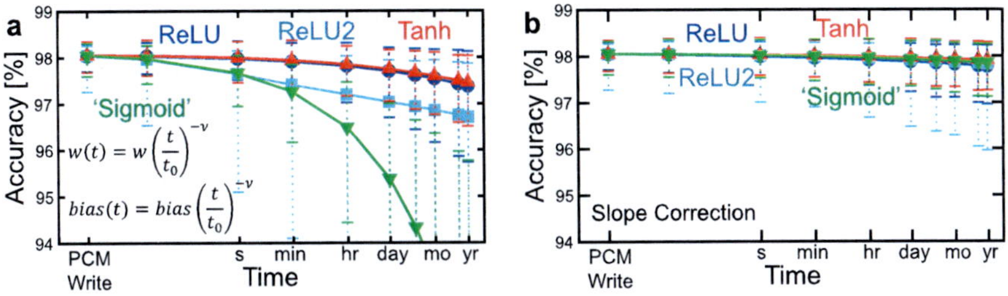

Fig. 17 **a** With no slope correction, even a small MLP suffers a strong accuracy loss from conductance drift. **b** However, with slope correction, accuracy barely degrades over time. The remaining decay is due to the σ_ν spread. Adapted with permission from [31]. Copyright 2019 IEEE

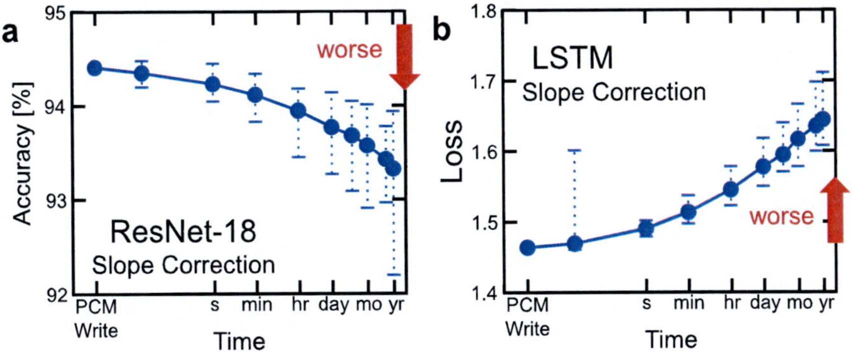

Fig. 18 On deeper, more sophisticated networks, the results still hold and slope correction is vital for maintaining accuracy over time for both ResNet and LSTM. LSTM results are measured according to the loss, where a lower loss is better, as a higher accuracy for ResNet is better. Adapted with permission from [31]. Copyright 2019 IEEE

more complicated networks, without any slope correction, accuracy was found to dramatically drop. However, slope correction is still highly effective in reducing the impact of conductance drift, as shown in Fig. 18. It is important to note that slope correction will not remove the entire impact of drift since the σ_ν is non-zero, meaning not all PCM devices drift at the same rate. However, even so, slope correction still shows a large impact, while remaining accuracy degradation can be compensated by using slightly larger networks [31].

3 Conclusion

In summary, analog hardware accelerators with phase change memory are a promising alternative to GPUs for neural network inference. The multiply-accumulate operation, which is the most computationally expensive operation in a neural network, can be performed at the location of the data, saving power and time. However, with this promise comes some non-idealities. Recent software-equivalent inference results [23, 29, 30] include strategies to program the weights more precisely using 4 PCM devices, and a slope correction technique can be used to reduce the impact of resistance drift [31].

References

1. Y. LeCun et al., Nature **521** (2015)
2. I. Goodfellow, Y. Bengio, A. Courville, *Deep Learning* (MIT Press, 2016)
3. V. Sze, Y. Chen, T. Yang, J.S. Emer, Proc. IEEE **105**, 2295–2329 (2017)
4. S. Hochreiter, J. Schmidhuber, Neural Comput. **9**, 1735–1780 (1997)
5. J. Schmidhuber, Neural Netw. **61**, 85–117 (2015)
6. Y. LeCun, L. Bottou, Y. Bengio, P. Haffner, Proc. IEEE **86**, 2278–2324 (1998)
7. I. Boybat, C. di Nolfo, S. Ambrogio, M. Bodini, N.C.P. Farinha, R.M. Shelby, P. Narayanan, S. Sidler, H. Tsai, Y. Leblebici, G.W. Burr, Improved deep neural network hardware-accelerators based on non-volatile-memory: the local gains technique, in *IEEE International Conference on Rebooting Computing (ICRC)* (2017)
8. J. Deng, W. Dong, R. Socher, L.J. Li, K. Li, L. Fei-Fei, ImageNet: a large-scale hierarchical image database. In: *CVPR09* (2009)
9. A. Krizhevsky, I. Sutskever, G.E. Hinton, Imagenet classification with deep convolutional neural networks. In: *Advances in Neural Information Processing Systems*, pp. 1097–1105I (2012)
10. B. Fleischer, S. Shukla, M. Ziegler, J. Silberman, J. Oh, V. Srinivasan, J. Choi, S. Mueller, A. Agrawal, T. Babinsky, N. Cao, C. Chen, P. Chuang, T. Fox, G. Gristede, M. Guillorn, H. Haynie, M. Klaiber, D. Lee, S. Lo, G. Maier, M. Scheuermann, S. Venkataramani, C. Vezyrtzis, N. Wang, F. Yee, C. Zhou, P. Lu, B. Curran, L. Chang, K. Gopalakrishnan, A scalable multi-teraops deep learning processor core for ai trainina and inference, in *IEEE Symposium on VLSI Circuits*, pp. 35–36 (2018)
11. J.L. Hennessy, D. Patterson, *Morgan Kaufmann* (1989)
12. N.P. Jouppi, C. Young, N. Patil, D. Patterson, G. Agrawal, R. Bajwa, S. Bates, S. Bhatia, N. Boden, A. Borchers, et al., In-datacenter performance analysis of a tensor processing unit, in *Proceedings of the 44th Annual International Symposium on Computer Architecture*, pp. 1–12 (ACM, 2017)
13. G.W. Burr, R.M. Shelby, S. Sidler, C. di Nolfo, J. Jang, I. Boybat, R.S. Shenoy, P. Narayanan, K. Virwani, E.U. Giacometti, B. Kurdi, H. Hwang, **62**, 3498–3507 (2015)
14. S. Ambrogio, P. Narayanan, H. Tsai, R.M. Shelby, I. Boybat, C. di Nolfo, S. Sidler, M. Giordano, M. Bodini, N.C.P. Farinha, B. Killeen, C. Cheng, Y. Jaoudi, G.W. Burr, Nature **558**, 60 (2018)
15. G.W. Burr, R.M. Shelby, A. Sebastian, S. Kim, S. Kim, S. Sidler, K. Virwani, M. Ishii, P. Narayanan, A. Fumarola, L.L. Sanches, I. Boybat, M.L. Gallo, K. Moon, J. Woo, H. Hwang, Y. Leblebici, Adv. Phys. X **2**, 89–124 (2017)
16. P. Narayanan, A. Fumarola, L. Sanches, S. Lewis, K. Hosokawa, R.M. Shelby, G.W. Burr, IBM J. (Res, Dev, 2017) 61 (2017)
17. I. Boybat, M.L. Gallo, R.S. Nandakumar, T. Moraitis, T. Parnell, T. Tuma, B. Rajendran, Y. Leblebici, A. Sebastian, E. Eleftheriou, Nat. Commun. **9**2514 (2018)
18. R.S. Nandakumar, M.L. Gallo, I. Boybat, B. Rajendran, A. Sebastian, E. Eleftheriou, in IEEE ISCAS Proc, pp. 1–5 (2018)
19. M.L. Gallo, A. Sebastian, R. Mathis, M. Manica, H. Giefers, T. Tuma, C. Bekas, A. Curioni, E. Eleftheriou, Nat. Electron. **1**, 246–253 (2018)
20. H.Y. Tsai, S. Ambrogio, P. Narayanan, R. Shelby, G.W. Burr, J. Phys. D Appl. Phys. **51**(2018)
21. S. Raoux, G.W. Burr, M.J. Breitwisch, C.T. Rettner, Y.C. Chen, R.M. Shelby, M. Salinga, D. Krebs, S.H. Chen, H.L. Lung, C.H. Lam, IBM J. Res. Dev. **52**, 465–479 (2008)

22. G.W. Burr, M.J. BrightSky, A. Sebastian, H.Y. Cheng, J.Y. Wu, S. Kim, N.E. Sosa, N. Papandreou, H.L. Lung, H. Pozidis, E. Eleftheriou, C.H. Lam, IEEE J. Emerg. Sel. Topics Circuits Syst. **6**, 146–162 (2016)
23. C. Mackin, H. Tsai, S. Ambrogio, P. Narayanan, A. Chen, G.W. Burr, Adv. Electron. Mater. **5**, 1900026 (2019)
24. S. Gupta, A. Agrawal, K. Gopalakrishnan, P. Narayanan, Deep learning with limited numerical precision, in *Proceedings of the 32nd International Conference on Machine Learning (ICML-15)*, pp. 1737–1746 (2015)
25. K. Spoon, S. Ambrogio, P. Narayanan, H. Tsai, C. Mackin, A. Chen, A. Fasoli, A. Friz, G.W. Burr, Accelerating deep neural networks with analog memory devices, in *IEEE International Memory Workshop (IMW)*, pp. 1–4 (2020)
26. A. Krizhevsky, V. Nair, G. Hinton, http://www.cs.toronto.edu/kriz/cifar.html
27. K. He, X. Zhang, S. Ren, J. Sun, Preprint 1512.03385 CoRR (2015) **abs/1512.03385** http://arxiv.org/abs/1512.03385
28. M. Boniardi, D. Ielmini, S. Lavizzari, A.L. Lacaita, A. Redaelli, A. Pirovano, IEEE Trans. Electron Devices **57**, 2690–2696 (2010)
29. H. Tsai, S. Ambrogio, C. Mackin, P. Narayanan, R.M. Shelby, K. Rocki, A. Chen, G.W. Burr, *2019 Symposium on VLSI Technology*, pp. T82–T83 (2019)
30. G. Cristiano, M. Giordano, S. Ambrogio, L. Romero, C. Cheng, P. Narayanan, H. Tsai, R.M. Shelby, G. Burr, J. Appl. Phys. **124**, 151901 (2018)
31. S. Ambrogio, M. Gallot, K. Spoon, H. Tsai, C. Mackin, M. Wesson, S. Kariyappa, P. Narayanan, C. Liu, A. Kumar, A. Chen, G.W. Burr, Reducing the impact of phase-change memory conductance drift on the inference of large-scale hardware neural networks, in *2019 IEEE International Electron Devices Meeting (IEDM)*, pp 6.1.1–6.1.4 (2019)
32. V. Joshi, M.L. Gallo, S. Haefeli, I. Boybat, S.R. Nandakumar, C. Piveteau, M. Dazzi, B. Rajendran, A. Sebastian, E. Eleftheriou, Nat. Commun. 2473 (2020)

Analogue In-Memory Computing with Resistive Switching Memories

Giacomo Pedretti and Daniele Ielmini

Abstract In the era of pervasive artificial intelligence (AI) and internet of things (IoT), achieving a high energy efficiency is at the top of priority for computing systems. In this scenario, in-memory computing is gaining momentum as a new methodology to overcome the von Neumann architecture and the related memory bottleneck. One of the most promising device for in-memory computation is the resistive switching memory (RRAM), also known as memristor, thanks to controllable conductance, good scaling and relatively low energy consumption. However, to achieve the promised benefits of in-memory computing with RRAM in terms of performance and power consumption, it is necessary to address a number of open challenges at the device, architecture and algorithm levels. This chapter presents the status of in-memory computing with RRAM, including the device concept and characteristics, the computing architectures and the applications. The perspective of analogue computing is analyzed with reference to both matrix vector multiplication (MVM) and inverse MVM to accelerate linear algebra problems that are generally executed with iteration schemes, highlighting the advantages in terms of performance, energy consumption and computational complexity.

1 Introduction

The computing industry has been always driven by an urge for an exponential growth. The microprocessor performance has increased substantially in the last 50 years thanks to aggressive scaling in the transistor channel size which led to a doubling of the number of transistors per square mm every 18 month, as predicted by the Moore's Law [1]. However, this scaling trend has been slowing down due to technological, physical and process related issues [2]. On the other hand, a similar exponential law is emerging in the recent years, namely the performance (measured in FLOPS, or Floating Point Operations per Second), required by artificial intelligence (AI) and

G. Pedretti · D. Ielmini (✉)
Dipartimento di Elettronica, Informazione e Bioingegneria, and IU.NET, Politecnico di Milano, Piazza L. da Vinci 32–20133, Milano, Italy
e-mail: daniele.ielmini@polimi.it

R. Micheloni and C. Zambelli (eds.), *Machine Learning and Non-volatile Memories*,
https://doi.org/10.1007/978-3-031-03841-9_4

scientific computing which have been observed to double every 3.4 months [3]. It is then clear that, on the one hand, new devices are needed for continuing with the Moore's Law trend, while from the other one an architectural design effort is desired to keep the pace of the required performance of modern AI algorithms.

From the architectural standpoint, the von Neumann architecture [4], which constitutes the mainstream architecture of most digital computers, suffers from the memory bottleneck. In fact, the memory and the central processing unit (CPU) are physically separated, thus most of the time is spent for data movement between memory and processing chips [5–7]. On the opposite, in-memory computing aims at executing all operations within the memory chip without any need for data movement [6, 7]. A promising memory technology for in-memory computing is the class of the resistive memory devices [8], often dubbed memristors [9, 10], featuring low energy operation, high speed, high density and compatibility with the complementary-metal-oxide-semiconductor (CMOS) technology [8, 11]. In-memory computing concepts based on resistive memories have been demonstrated with several memory technologies, including the resistive switching random access memory (RRAM), the phase change memory (PCM), the ferroelectric random access memory (FERAM) [12] and the magnetic random access memory (MRAM) [8, 13]. Different computing concept can be executed inside the memory such as logic computing [7, 14], neuromorphic computing [15–19], stochastic operations [20] and analog computing [21]. In particular, analogue computing allows to accelerate several computing operations thanks to physical computation, where multiplication and summation are executed by physical laws in the analogue domain. Also, the unique architecture of the crosspoint memory array allows to parallelize computing, thus enabling a reduction of computational complexity with respect to the conventional digital computing. At the same time, in-memory computing is prone to errors and inaccuracies due to noise and device variations, which should be carefully taken into account for a fair comparison with the floating-point precision in digital circuits.

This chapter aims at reviewing the recent advances of analog in-memory computing with RRAM devices. First, we present the device properties and characteristics, in particular discussing the device requirements for analogue memory. Then we present the main analogue computing architectures with RRAM, focusing on matrix-vector-multiplication (MVM) for neural networks and optimization algorithms. Finally, we present the inverse-matrix-vector-multiplication (IMVM) architecture and illustrate the main applications and their advantages and drawbacks in terms of energy efficiency, time complexity and precision.

2 Resistive Switching Memories

Various types of nanoelectronic devices have been proposed in the latest years to replace or complement the conventional CMOS memory technologies at various levels of the hierarchy. Most of these memories rely on the concept of changing

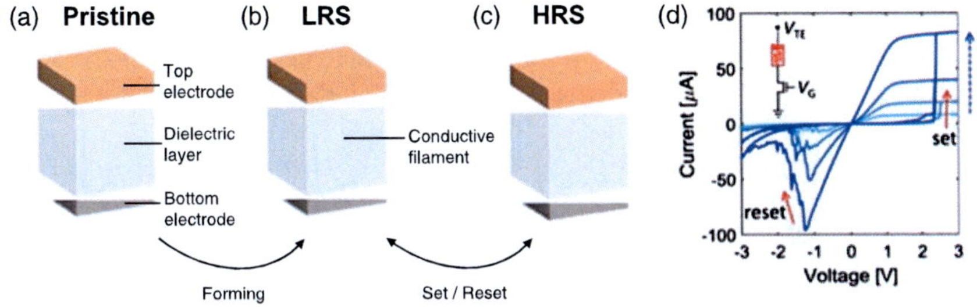

Fig. 1 RRAM devices. **a** RRAM are made by a dielectric material inserted between two metallic top electrode (TE) and bottom electrode (BE). In the pristine state they show a high resistance state (HRS). **b** By applying a positive forming voltage at the TE a filament grows from TE to BE resulting in low resistance state (LRS). **c** To retrieve the HRS it is possible to apply a reset voltage. **d** Typical I-V curve of a RRAM device in 1T1R configuration showing the ability of analog programming through different compliance current (or gate voltage V_G) and different reset voltages. Reprinted from [31] under Creative Commons License

the active material properties by the application of voltage or current programming pulses, thus storing the memory states as a specific material configuration. Several technologies for two terminal devices, such as resistive switching memory (RRAM) [11, 22], PCM [23, 24], FERAM [12, 25, 26] and MRAM [27], have been proposed. Among them, RRAM have attracted widespread research interest thanks to its low energy [28], high speed [29] and high density, combined with the ability of 3-dimensional integration [30]. Figure 1 shows the RRAM device structure, operation and switching characteristics. Typically, the device consists of a metal–insulator-metal (MIM) structure in its pristine state (Fig. 1a), which is a high resistance due the dielectric insulating layer. The device is generally initialized by the forming operation, consisting of the application of a relatively high voltage between the top electrode (TE) and bottom electrode (BE). During the forming process, the device undergoes a soft breakdown event with a local variation of the material composition, consisting of a low-resistivity filament which is responsible for the low resistance state (LRS, Fig. 1b). Then, it is possible to recover a high resistance state (HRS) by the application of a reset negative voltage between TE and BE which forms a depleted gap in the filament (Fig. 1c) that can be controlled by the maximum applied negative voltage. A set positive voltage pulse causes the device to switch back to the LRS with a continuous filament connecting the TE to the BE. The filament size is generally controlled by the maximum current flowing during the set operation, known as compliance current I_C and usually regulated by a series transistor. Figure 1d shows a typical current–voltage (I–V) curve of a RRAM device in a one-transistor-one-resistor (1T1R) structure with HfO_2 switching layer [31], highlighting the various states obtained by the modulation of the compliance current, which depends on the gate voltage V_G of the series transistor. Note that different resistive states are achieved at increasing I_C, thus demonstrating that RRAM can be used not only as a digital memory storing a ‘1’ in the LRS and a ‘0’ in the HRS, but also as a continuous analog memory with multiple states corresponding to different resistive values. The first advantage is that multiple levels, hence multiple bits, can be stored in a single memory element, thus increasing the bit density in a memory array. Secondly, novel computing applications harnessing the analogue tuning of RRAM device can be unleashed to perform analogue in-memory computing [7, 21].

2.1 Memory Array Structures

RRAM devices can be arranged in various memory array structures for both as memory and computational unit, as shown in Fig. 2. The most straightforward configuration is passive crosspoint array where the RRAM device displays a simple one-resistor (1R) structure (Fig. 2a). In the crosspoint array each RRAM device is located at the intersection between a row line and a column line connecting the BE and TE of the device, respectively [32] By programming a conductance G_{ij} in the RRAM device connected between row i and column j of the crosspoint array and applying an input voltage vector $V = (V_1, V_2, \ldots, V_N)$ at the column terminals by keeping

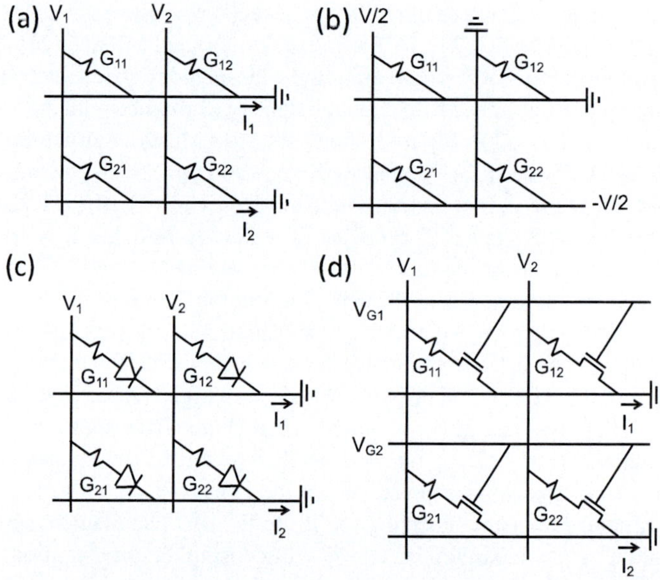

Fig. 2 Memory structures. **a** 1R crosspoint array where every device with conductance G_{ij} is connected between each row and column. **b** $V/2$ programming scheme in 1R crosspoint array, where only the selected cell (blue) receive the whole V voltage necessary for programming while the undesired selected cells (red) receive only $V/2$. **c** 1S1R crosspoint array, where a two terminal select device is inserted in series with every RRAM to mitigate the sneakpaths problem. **d** 1T1R array, where a transitory is used as select device and to regulate the compliance current during the set operation enabling analog programming. Reprinted with permission from [21] under Creative Commons License

the rows at ground, the resulting current is given by the matrix vector multiplication (MVM) formula:

$$I_i = \sum_{j=1}^{N} G_{ij} V_j \tag{1}$$

where N is the size of the crosspoint array. Note that the MVM operation naturally arises in the crosspoint array thanks to physical laws, namely the Ohm's law for the multiplication $I = GV$ and the Kirchhoff's law for the summation of individual currents incurring at the same row. Since MVM is a basic algebra operation, the crosspoint array structure is widely using in most analogue in-memory computing applications [7, 21, 33, 34]. The crosspoint array structure also takes advantage of a high memory density due to a memory cell area occupation of only $4F^2$, where F is the lithographical feature. Array organization in 3D arrays usually results in even smaller cell effective area, which is highly favorable for computing with large amounts of data [35]. However, the crosspoint array structure has the strong drawback of the difficult programmability and read disturb induced by sneak-path effect [36]. In fact, while selecting cell G_{ij} for set, reset or read operation, the cell row i and column j are biased, which results in unwanted current flows even if the unselected terminals are left floating, resulting in read disturb or possible set or reset operations at the unselected devices. Disturbs and sneak paths can be mitigated by suitable bias schemes, such as the $V/2$ biasing scheme in Fig. 2b [37, 38], where a voltage $V/2$ is applied to column j and $-V/2$ is applied to row i with all other rows/columns grounded. As a result, the voltage across all unselected cells is zero, except for the half-selected cells along row i and column j, where the voltage is reduced by a factor 2 thus minimizing the probability of undesired set or reset events. During the read operation, a voltage V_R is applied to the selected column while all the other columns and rows are connected to ground, which allows for reading all cells of the selected column in parallel [38].

However, due to the strong variation of set and reset voltages and to the limited set/reset resistance window, the passive crosspoint array with 1R structure can only be used with small array size, while becoming unpractical for the most typical array size for memory and computation.

2.1.1 1S1R Structure

To enable large crosspoint array size, a two terminal select device should be add, resulting in a one-selector/one-resistor structure (1S1R) [39–41] as shown in Fig. 2c. Selector devices should display a strong non-linear characteristic to prevent any current flowing in half selected devices with $V/2 < V_t$, where V_t is the selector device threshold voltage. Also, the select device should display large current at relatively large voltages to enable set and reset processes within the selected device.

The nonlinear characteristic should also be bidirectional, i.e., operable at both positive and negative voltages for set and reset processes, respectively. 1S1R crosspoint arrays are extremely promising for memory application, in particular for storage class memories to fill the performance/cost gap between DRAM and Flash memory. On the other hand, in-memory computing applications of 1S1R structures are yet to be unveiled, the main challenge being the contribution of the non-linear selector to the MVM operation.

2.1.2 1T1R Structure

The RRAM device can be connected in series with a transistor selector resulting in a one-transistor/one-resistor (1T1R) structure, as shown in Fig. 2d. To select a memory cell within an array for a set operation, the corresponding transistor gate line should be biased to turn on the transistor enough such that the applied TE voltage developed across the selected RRAM exceeds the set voltage. The transistor can be used for controlling I_C during the set operation, thus tuning the filament size hence the RRAM conductance, which makes the 1T1R structure ideal for analogue programming [42, 43]. During the reset or read operation, the selected gate line should be biased at a high voltage, such that all the TE voltage applied across the selected RRAM drops across the device, since the transistor resistance is negligible. Due to the excellent control of analogue state and lack of sneak paths, the 1T1R structure is by far the preferred configuration to demonstrate analogue-type in-memory computing functions [34, 44]. For the same reasons, we will restrict our focus on 1T1R structures in the following.

2.2 Requirements for Analogue Memory

The 1T1R structure allows to gradually control the conductance both during the set operation (via I_C) and the reset operation (via V_{stop}). In fact, the multilevel capability is a key requirement for analogue computing, making the memory able of representing multiple states in a single cell. In principle, if the available memory have only a few (or even two) possible resistance states, it is possible to memorize different slices of the analogue information in multiple devices [45], but the area occupation increases significantly. Another important requirement is the linearity of the I–V curve, so that by applying increasing input voltages, the cell current response increases linearly, thus satisfying Ohm's law for analogue multiplication. Note that, in most applications, this requirement can be circumvented by applying the input as a train of digital pulses with a simple unary or a more compact shift and add [45] encoding, and then reconstruct the analogue output by properly integrating the current in the time domain.

In general, partial HRS configurations obtained by the reset operation tend to show a non-linear characteristic, due the non-ohmic conduction in the depleted gap

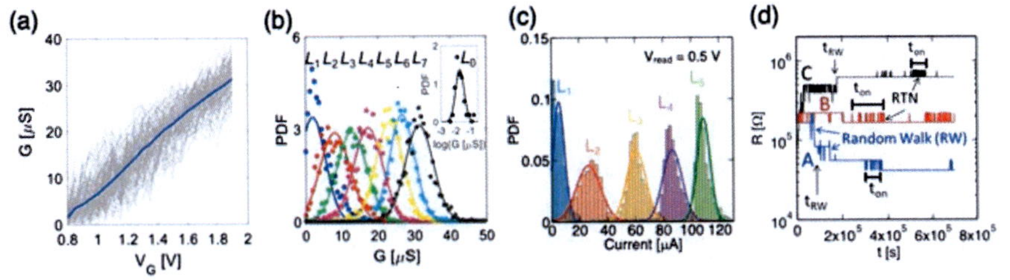

Fig. 3 Analog RRAM programming non-idealities. **a** Measured RRAM conductance as function of the gate voltage VG, showing an average linear dependence (blue), while single traces show large variations (grey). **b** Distribution of 7 levels of LRS obtained by modulating V_G and 1 level of HRS (inset) showing a conductance independent standard deviation. **c** Five analog levels programmed in a 4 kB RRAM array showing the both cell-to-cell and cycle-to-cycle variability. **d** Fluctuations of a programmed state on a RRAM device in HRS, showing different phenomena such as random walk and random telegraph noise. Reprinted with permission from [31, 43] under Creative Commons License. Reprinted with permission from [47]. Copyright 2015 IEEE

along the filament [11]. For this reason, it is most common to adopt I_C-controlled LRS configurations by set operation for preparing analogue states with variable conductance. Figure 3a shows the conductance as function of gate voltage in a 1T1R structure for 100 cycles and the median value [31]. At every cycle, the device was first prepared in the HRS, then a train of set pulses with fixed TE voltage $V_{TE} = 3$ V above the threshold for set voltage and increasing gate voltage V_G was applied. As expected, the median value increases linearly with $V_G - V_T$, where $V_T = 0.7V$ is the transistor threshold voltage. However, one cannot solely rely on V_G (or equivalently I_C) for precisely controlling the device in a desired conductance, due to the cycle-to-cycle variations of the traces in Fig. 3a. These variations are generally attributed to the stochastic ionic migration during the set operation, which leads to variations in the shape and volume of the conductive filament [46, 47]. Figure 3b shows the resulting Gaussian distributions of conductance for 7 programmed LRS levels, indicating a standard deviation σ_G=3.8 μS. In addition to the cycle-to-cycle variability, a device-to-device variability arises as different devices in an array usually present different characteristics due to variation in the fabrication process causing specific geometry and material composition within the RRAM cell. Figure 3c shows distributions of currents for a read voltage $V_{read} = 0.5V$ of 4 analogue levels programmed on a 4 kB RRAM array [43]. The observed variation includes contribution due both to cycle-to-cycle and device-to-device variability. To mitigate both cycle-to-cycle and device-to-device variability, it is possible to adopt program and verify algorithm. For instance, given a certain conductance G_{target} that should be approximately reached in the RRAM device, one can gradually increase the gate voltage as shown in Fig. 3b until the conductance G reaches a value between $G_{target} - G_{tol}$ and $G_{target} + G_{tol}$, where G_{tol} is the acceptable tolerance. If G exceeds $G_{target} + G_{tol}$, then a reset pulse can be applied to reduce G within the acceptable range. If G goes below $G_{target} - G_{tol}$ then another set pulse can be applied, until convergence into the [48, 49]. In principle, program-verify techniques allow to reach any desired conductance states within an arbitrary tolerance range at the cost of increasing circuit complexity, programming energy and time. Also program-verify techniques tend to result in accelerated wear out of the memory device. In fact, after multiple set-reset operations the RRAM can be found in a non-ideal state, such as a stuck-off or stuck-on state [50]. One should carefully tune the maximum number of iterations during a program and verify routine to balance precision and device degradation.

After the RRAM device is programmed, the conductance state is also prone to time-dependent variations which may lead to conductance G to drift out of the tolerance range. In fact, RRAM suffers from resistance fluctuations over time, as shown in Fig. 3d for a HfO_x RRAM device [51]. The RRAM initially programmed at a given resistance might either increase or decrease its value, due to intermittent random telegraph noise (RTN) and random walk, which makes deterministic analog programming extremely challenging.

3 In-Memory Computing Architecture for Matrix-Vector Multiplication

The RRAM crosspoint array allows to execute the MVM operation simultaneously, in one step and in the analogue domain, thanks to physical Ohm's law multiplication and Kirchhoff's law summation of currents in (1) [33]. Figure 4a illustrates the basic MVM operation while Fig. 4b shows the correlation plot of the measured output currents as a function of the ideal MVM results obtained for a crosspoint array [48]. Motivated by the ubiquitous importance of MVM operations in data analytics and computing workloads, RRAM crosspoint for analogue MVM acceleration have been demonstrated for multiple applications [21, 32], such as neural networks acceleration [34, 52–55], image processing [44, 56], optimization algorithms [57–60], hardware

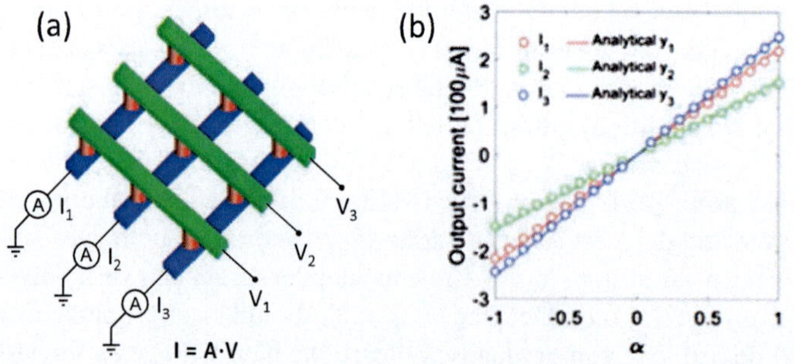

Fig. 4 MVM in crosspoint arrays. **a** A crosspoint array can be used for accelerating MVM. By programming a matrix A into the crosspoint conductance and applying a voltage vector V on the columns, the resulting current flowing into the rows tied to ground is $I = AV$. **b** MVM output current as function of α given an input voltage $V = \alpha V_0$. Reprinted with permission from [48], under Creative Commons License

security [38, 61, 62] and accelerated solution of differential equations [63]. Integrated circuit comprising CMOS mixed-signal circuits and RRAM arrays fabricated in the back end of the line have already been realized for several applications [64–67]. The most promising applications of MVM are probably neural network and optimization acceleration.

3.1 *In-Memory Neural Network Accelerators*

Analogue in-memory MVM has been widely used for feed-forward operation in neural network acceleration [21]. Figure 5a shows a conceptual illustration of a three-layer perceptron neural network [68]. Input data are applied on the left side, evaluated by the network layers from left to right, until reaching the output layer. At each layer in this forward transition, every neuron n_i emits a signal x_i which is multiplied by the synaptic weight w_{ij} before reaching the output neuron m_j. The evaluation of the output state y_j corresponding to neuron m_j consists of the summation of all the contributions from the previous layer, according to the formula $y_j = \sum_i x_i w_{ij}$, which is analogous to (1) where y_j is replaced by a physical output current, x_i is an input voltage and w_{ij} is the RRAM conductance. The forward operation of neural networks can thus be evaluated by an analogue MVM operation within a crosspoint array, with a throughput improvement up to 10^4 compared with multiply-accumulate (MAC) operations executed on a traditional digital computer [69]. Note that a RRAM device can only store positive weights whereas w_{ij} generally comprise both positive and negative values. Figure 5b–c show two possible techniques to enable mapping relative numbers as weights in a crosspoint neural network. In general, two RRAM devices can be used in parallel, each being biased with opposite polarity voltages to represent both negative and positive weights. For instance, a reference fixed conductance G_{ref}

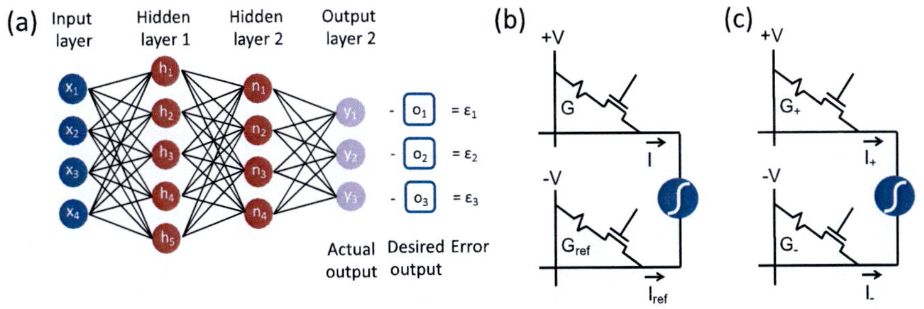

Fig. 5 Neural networks weights implementation. **a** Multilayer perceptron with an input layer, 2 hidden layers and an output layer. The output y is compared with the ground truth o and the calculated error is backpropagated for training the network. **b** representation of positive and negative weights with a single programmable RRAM and a fixed reference conductance. **c** a more flexible bipolar weight representation with two programmable RRAM devices. Reprinted with permission from [21] under Creative Commons License

with applied negative bias can be used in parallel to a programmable RRAM with conductance G_+ with applied positive bias such that the equivalent conductance is given by $G = G_+ - G_{\text{ref}}$, thus the modulation of G_+ allows to obtain both positive and negative weights, as shown in Fig. 5(b). A more flexible and granular approach is to use two programmable RRAM devices biased with opposite polarities with conductance G_+ and G_- to represent the overall weights as $G = G_+ - G_-$, as shown in Fig. 5(c). In this way, it is possible to program independently both conductance thus increasing the number of programmable weights in the 2-RRAM structure [70, 71].

In-memory computing can not only accelerate the feed-forward processing, also known as the inference phase, but also the training phase of neural networks [52, 53, 64]. In this case, after the forward evaluation of a single or multiple elements of a dataset, the weights are updated based on a learning rule [68]. A typical supervised training algorithm is backpropagation, where the output state of a neuron y_j is compared with its ideal result o_j and an error $\varepsilon_j = y_j - o_j$ is computed. This error is then back-propagated to the weights that are updated by an amount $\Delta w_{ij} = \eta x_i \varepsilon_j$, where η is the learning rate that controls the speed on which weights are updated and can be an important parameter for controlling convergence and overfitting. In the training operation, to compete with conventional digital hardware, the weight update should be both fast and precise [72], thus the requirements for computational memory are more aggressive. To enable both fast and precise training, an important feature is the linearity of the weight update [73].

The device characterization procedure to demonstrate the feasibility of online network training usually consists of the application of a train of programming pulses with a constant amplitude and shape for the increase and decrease of conductance. The ideal expected result is shown in Fig. 6a, where the weight value as a function of the number of programming pulse increases linearly under applied positive voltage pulses until reaching a maximum value (i.e., 1 on the relative axis of the figure), then returns to 0 under applied negative voltage pulses. The weight update ΔG should be independent of the starting conductance value, thus allowing the weight update even without reading the initial conductance thus speeding up the training process. As shown in Fig. 6b, RRAM devices generally show non-linear potentiation (increase of G) and depression (decrease of G), where the set/reset pulses have an abrupt effect of the conductance change followed by a saturation after a few pulses [71]. RRAM may also have an asymmetrical weight update, as shown in Fig. 6(c), due to different update rates in the potentiation and depression processes. Asymmetric update translates in a larger number of pulses needed for a positive update ΔG than a negative update $-\Delta G$, or vice versa. Usually, there is a characteristic conductance value G_{sym} where the positive and negative increments have the same amplitude [74]. In this case, it is possible to use the scheme of Fig. 5(b) with $G_{\text{ref}} = G_{\text{sym}}$, so that a symmetric potentiation/depression response is obtained [74, 75].

RRAM devices also usually display a limited conductance window spanning from $G_{\min} > 0$ to $G_{\max} < 1$ where the device can be programmed. This result is an offset from the ideal case as shown in Fig. 6d. In such a case, the weight configuration of Fig. 5b can help reaching the desired conductance by carefully tuning G_+ and G_-.

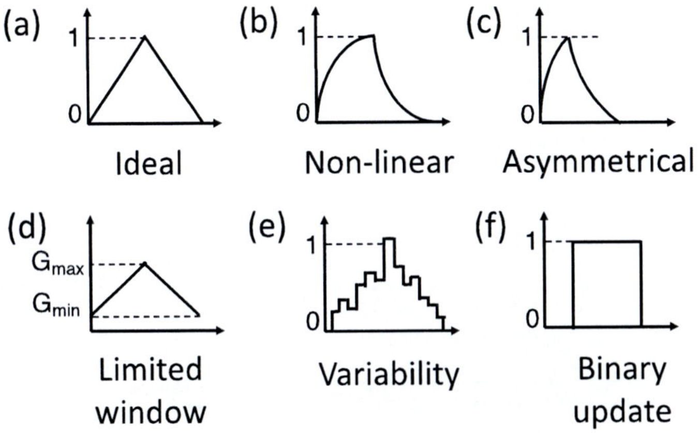

Fig. 6 Weight update characteristics. **a** Ideal weight update characteristics, with the conductance G linearly increasing and decreasing with positive and negative voltages pulses, respectively. **b** Non-linear weight update characteristic with G that after a steep increase saturates. **c** Asymmetrical weight update, with a different response to positive and negative pulses. **d** weight update characteristic with a limited conductance window. **e** Variability in weight update due to cycle-to-cycle variability. **f** Weight update with binary device. Reprinted with permission from [21] under Creative Commons License

In particular, programming the same conductance in G_+ and G_- devices allows to reproduce the case $w_{ij} = 0$, which is among the most probable values within the distribution of synapses in typical neural networks. As already discussed, RRAM also shows stochastic variations in the set and reset processes, which can lead to an unpredictable weight change during training as shown in Fig. 6e, thus affecting significantly the convergence operation. Stuck-on and -off states, where the conductance cannot be updated, further complicates the scenario.

In an extreme case RRAM devices could show only two conductance states (HRS and LRS) [76, 77] resulting in a binary weight update scheme as shown in Fig. 6f. This makes the weight encoding inherent digital which is suitable for the acceleration of a binary neural network (BNN). Training a BNN can be challenging due to the abrupt change of conductance. Figure 7a illustrates a stochastic approach for training BNNs using binary RRAM devices with an internal parameter controllable with the application of multiple programming pulses [76]. Two different weights value can be associated with the RRAM device, namely W_{int} and W_{ext} corresponding to a non-observable internal variable and the externally measured weight, respectively. W_{int} may correspond to the defect density and filament configuration within the device while W_{ext} corresponds to the measured conductance, as shown in Fig. 7a. The binary weight W_{ext} can only assume two values, 0 if the defects do not connect TE and BE, and 1 otherwise. By the application of multiple set and reset pulses it is possible to change the filament configuration, hence the continuous update of W_{int}, while W_{ext} only changes after a certain threshold is reached. This hybrid binary/analogue update can be adopted within a conventional backpropagation algorithm for training

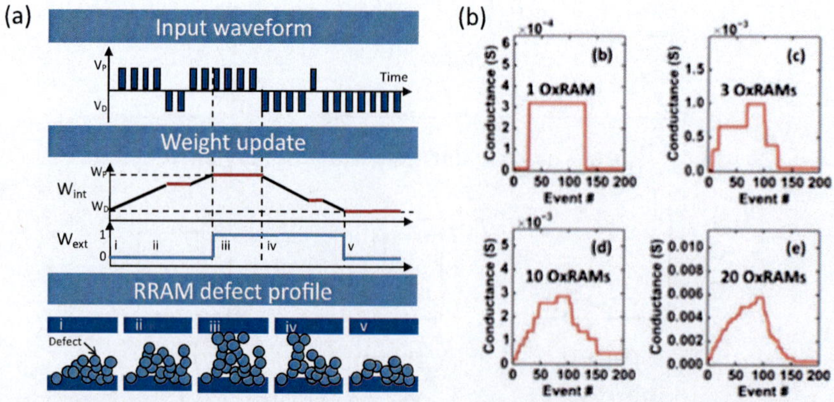

Fig. 7 Using binary devices for training neural networks. **a** Stochastic weight updated, where the state variable is represented by an internal weights which is updated by positive/negative pulses (top) and it is not measurable, and an external binary weight which is updated after multiple positive/negative pulses (center). The internal weight update represents a modification in the defect configuration while the external weight changes if a connection is created or broken from TE to BE (bottom). **b** Multiple binary devices used as synapse to represent an analog weight. Reprinted with permission from [76, 78]. Copyright 2017, 2015 IEEE

a BNN as if it was an analogue neural network. Similarly, W_{ext} can be calculated after measuring a W_{int} value, making it possible to use an analogue memory for training a BNN which usually shows high precision [77]. Another approach is illustrated in Fig. 7b, where multiple binary RRAM devices are used for training a conventional neural network [78]. In this procedure, multiple devices are connected in parallel to represent multiple weights. Every time an individual binary RRAM device increases/decreases its conductance, the overall weight experiences a corresponding incremental step. As the number of devices increases, the synaptic weight becomes increasingly analogue, thus making the characteristic similar to the ideal one. This comes at the cost of an increased area occupation. However, multiple devices in parallel can also be used to mitigate the effect of non-idealities and stochastic weight update [79] thus serving as a regularization of individual variations to achieve a more gradual weight update response in the presence of particularly unprecise devices.

RRAM devices have also been shown to be able of brain-inspired spike-based neural network implementation [18, 19, 80, 81]. In this case, information is usually encoded in spikes similarly to our brain, where learning takes place according to unsupervised weight update rules depending on the spike timing, such as the spike timing dependent plasticity (STDP) [18], the spike-rate dependent plasticity (SRDP) [80, 81] or semi-supervised training approaches implementing a teacher signal [19]. In most practical implementations, RRAM devices are used as artificial synapses, although fully-memristive architecture with RRAM-based synapses and neurons have also been presented [82].

3.2 In-Memory Optimization Accelerators

MVM is also at the core of many optimization algorithms, such as linear or quadratic programming techniques [83]. Specifically-designed neural networks can be implemented for searching the minimum of an energy landscape, usually relying on Hopfield neural networks (HNN) [84], namely recurrent neural networks where each neuron is connected to all the others with symmetric links ($w_{ij} = w_{ji}$) and no self-connection ($w_{ii} = 0$). This brain-inspired recurrent connectivity offers interesting cognitive functions, such as attractor learning/recall and associative memory [85], that have also been demonstrated with in-memory computing hardware [86–88].

HNNs also have shown the ability of solving constraint satisfaction problems (CSP) [89], which are ubiquitous in many different application fields [90]. In this case, every neuron has a highly nonlinear activation function and represents a state of the network, while connectivity between neurons define the constraints. By initializing a random input state, the network can gradually update its states and converge to an optimized final state by minimizing the energy landscape cost function given by:

$$E = -\frac{1}{2}\sum_{i,j}^{N} w_{ij} v_i v_j \tag{2}$$

where N is the total number of neurons and v_i represents the state of neuron i. The binary neuron state is updated depending on the evaluation of the input function $u_i = \sum_{i \neq j} w_{ij} v_j$ compared with a given threshold θ_i. Figure 8a shows an example of a convex energy cost function, which can be explored by the HNN to find the minimum, i.e. the optimization problem solution. Convex problems can be computed straightforwardly by conventional gradient descent techniques. However, when the the problem size and difficulty may quickly increase in typical CSP, which are known to become aggressively difficult as in the case of non-deterministic polynomial (NP) or NP-hard problems. This is due to the increase of the number of local minima in the energy landscape, where the the HNN state can be stuck as illustrated in Fig. 8b.

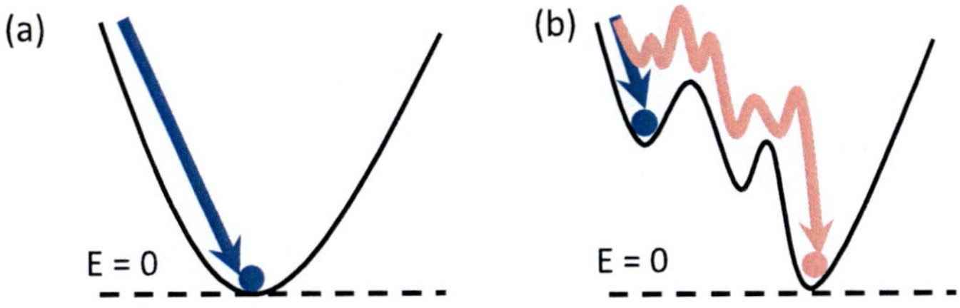

Fig. 8 Energy landscapes of optimization problems. **a** Search of the minimum of a convex energy landscape with a Hopfield neural network. The solution (blue) can reach it efficiently. **b** Search of the global minimum of a non-convex energy landscape with the deterministic solution (blue) being stuck in a local minimum. By adding noise (red) the solution can efficiently reach the global minimum

Non-convex CSPs can be solved by the simulated annealing technique [91]. By stimulating the HNN with random noise, it is possible to 'heat' the system, thus helping the state in escaping from the local minima and eventually reaching the correct solution in the global minimum of the energy function. Simulated annealing accelerators have been demonstrated by conventional CMOS circuits [92–95], quantum computing technologies [96, 97], optical computing technologies [98] and analogue in-memory computing [57–60, 99–102]. In the latter, memory devices can act both as MVM accelerators for the inference of the HNN and annealers by the generation of intrinsic random noise.

Figure 9a shows a conceptual circuit schematic of a HNN for accelerated annealing [59], which is based on an MVM current which is then fed back into the input neurons. The feedback is obtained by sampling the columns currents with an analogue-to-digital converter (ADC), post-processing it to obtain the neuron states and applying it to the crosspoint rows in either digital or analogue mode. The constraints are encoded in the weight matrix of the crosspoint conductance, while the intrinsic noise to stimulate the annealing can be generated by various techniques. For instance, one can leverage RRAM inherent stochasticity such as RTN [103] and 1/f noise [104], which can automatically stimulate the simulated annealing [102]. An additional crosspoint column can be used to program RRAM devices appropriately and then harvest noise

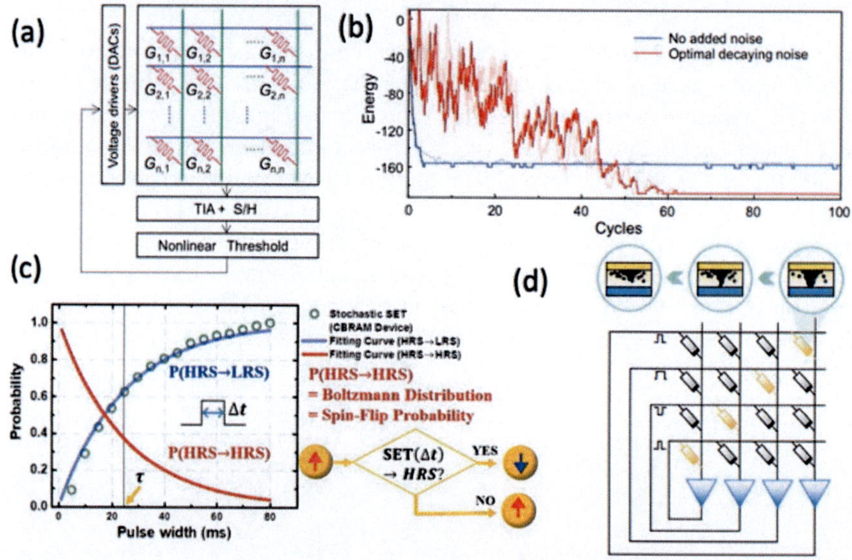

Fig. 9 In-memory simulated annealing techniques. **a** an extra column of a crosspoint array can be used to generate noise which is summed to the MVM response to perform simulated annealing. **b** Hopfield energy as function of iteration cycles for different noise levels. **c** stochastic switching of a RRAM device to generate random flip of a neuron state. The probability of switching can be finely tuned by regulating the set pulse width. **d** Hopfield network working in the chaotic regime by inserting diagonal connection that are gradually reduced in weight cooling the overall annealing procedure. Reprinted with permission from [58, 60, 66]. Copyright 2019, 2020 IEEE

with the desired figure to speed up optimization [59]. In fact, noise should be modulated based on the annealing scheme and should also change its characteristic during the annealing procedure, ideally reducing its magnitude while reaching the global minimum. Figure 9b shows the Hopfield energy to solve a 60-node Max-Cut problem as a function of iteration cycles for different noise levels [59]. A noise-free calculation results in a higher energy compared to noisy calculations, thus supporting the fundamental contribution of noise for the annealing. Also note that noise with large amplitude might be inefficient for reaching the energy minimum, since the state might also escape from the energy global minimum in this case. The tradeoff between exploratory and greedy strategies should be therefore carefully considered. A second approach is to use the RRAM stochasticity in its switching to generate a flip of one or more neurons with a given probability as shown in Fig. 9c [58]. In fact, the probability of setting a RRAM device can be controlled with the set pulse itself either by the voltage pulse amplitude [20] or the pulse width [58]. In this way, after a careful characterization of the set statistics of the RRAM device, it is possible to create naturally a stochastic, e.g., Gaussian, distribution from the device physics. A third approach is to operate the HNN in its chaotic behavior [105], by violating one of its definitions, namely $w_{ii} = 0$ by connecting each neuron in a self-feedback to itself as shown in Fig. 9d [57, 60]. By gradually resetting the self-feedback RRAM device the chaos can be reduced to let the system effectively reach the global minimum.

Optimizer circuits based on hardware HNN accelerated by RRAM crosspoint arrays have been shown to have better performance than traditional, optical and quantum computing [59], thanks to the low energy MVM operation and intrinsic, compact noise generators.

4 In-Memory Computing Architecture for Inverse MVM

MVM can be used to accelerate algebraic problems, such as the solution of linear systems and partial differential equations [63, 106]. However, this is usually done by iteratively performing the MVM operation, digitalize the currents with an ADC, post-process them and apply the correct output vector as input for the following MVM. While MVM can indeed accelerate the algebraic problem, solution compared with digital approaches, the number of iterations for reaching the convergence can be extremely large. To further accelerate the problem solution, the feedback operation can be obtained within the analog domain, by operational amplifiers connected between rows and columns of crosspoint arrays [31, 48, 107] as shown in Fig. 10a. Given a crosspoint array programmed with a conductance matrix G, by injecting a current vector I to its rows connected to the negative input of an operational amplifier (OA) with the positive input connected to ground, the columns connected to the OA outputs will adjust to a voltage V such that the overall current flowing within the OA is zero due to its high input impedance and negative feedback effect, namely $I + GV = 0$ [48]. This leads to:

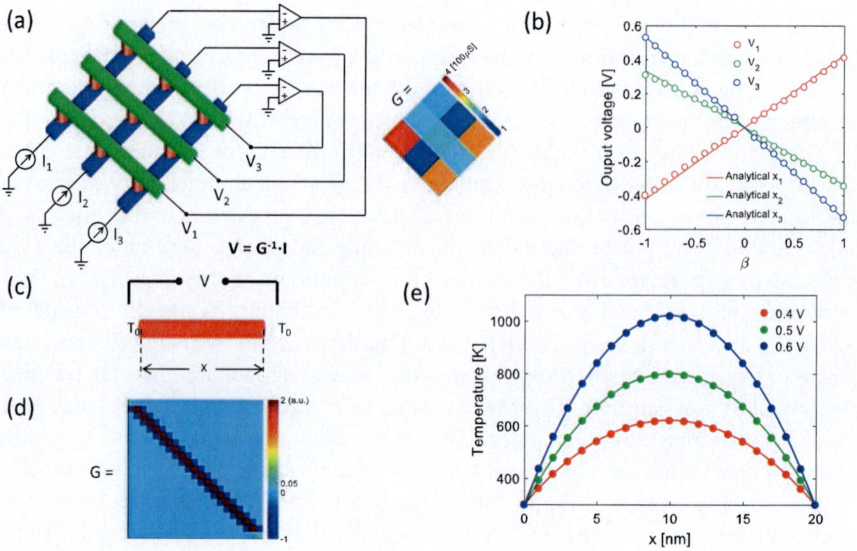

Fig. 10 In-memory solution of linear systems with IMVM. **a** IMVM circuit where a current is injected in a crosspoint programmed with conductance G rows which are connected to the virtual ground of operational amplifiers whose output is connected to the crosspoint columns. The result gives $V = G^{-1} I$. An example of measured 3×3 programmed matrix (inset). **b** Circuit output voltage as function of β with an input current $I = \beta I_0$. **c** representation of a 1-dimensional Fourier heat equation problem. **d** Fourier equation encoded in a crosspoint array. **e** Circuit simulation (circles) results compared with analytical solution showing good agreement. Reprinted with permission from [48] under Creative Commons License

$$V = -G^{-1} I \tag{3}$$

which corresponds to the solution of a linear system. In fact, by encoding in G a problem A and injecting a current $-I$ corresponding to a known term b, the resulting output voltage will be equal to $x = A^{-1}b$ as shown in Fig. 10a. This operation can also be referred as inverse MVM (IMVM), as the solution is the vector which must be multiplied to the given matrix to yield a certain output vector. Figure 10b shows an experimental demonstration of this circuit, where a 3×3 crosspoint array of RRAM devices was connected in feedback with OAs on a printed circuit board (PCB) [48]. By applying an input current vector with amplitude $I = \beta I_0$, where I_0 is a normalized vector and β represents the magnitude of the input vector, the measured voltage at the OA output displays a linear dependence on β as expected from the linearity of the system of equation, thus demonstrating the feasibility of the circuit in the solution of the linear system. Interestingly, the solution of a linear system is obtained in just one step by the circuit, without any iteration. Moreover, the time to solution does not depend on the matrix size, thus making the time complexity of the circuit constant, i.e., $O(1)$ complexity [108, 109]. This is extremely attractive in comparison with traditional algorithms such as conjugate gradient [110] or quantum computing algorithms such as the Harrow-Hassidim-Lloyd (HHL) algorithm [111], which show a time complexity of $O(N)$ and $O(\log(N))$, respectively. Figure 10c shows a real word problem, the solution of a 1-dimensional steady-state Fourier equation for heat diffusion encoded in a crosspoint array and solved by the IMVM circuit. Matrix G in Fig. 10d represents the system of linear equations which describes the differential Fourier equation in the discrete domain by the finite difference method (FDM). Note that G displays both positive and negative coefficients, which can be encoded with a method similar to Fig. 5c, where the output voltage of the OAs are inverted and applied to a second crossbar representing the negative entries [48]. The known term encoded in the input currents correspond to the dissipated power in the one-dimensional structure and the output voltage represents the temperature profile along the 1-dimensional structure. Figure 11e shows the results obtained by a SPICE simulation of the IMVM circuit compared with the analytical solution for different voltage applied to the 1-dimensional structure, highlighting a accurate match between the ideal analytical solution of the equation and the circuit simulations.

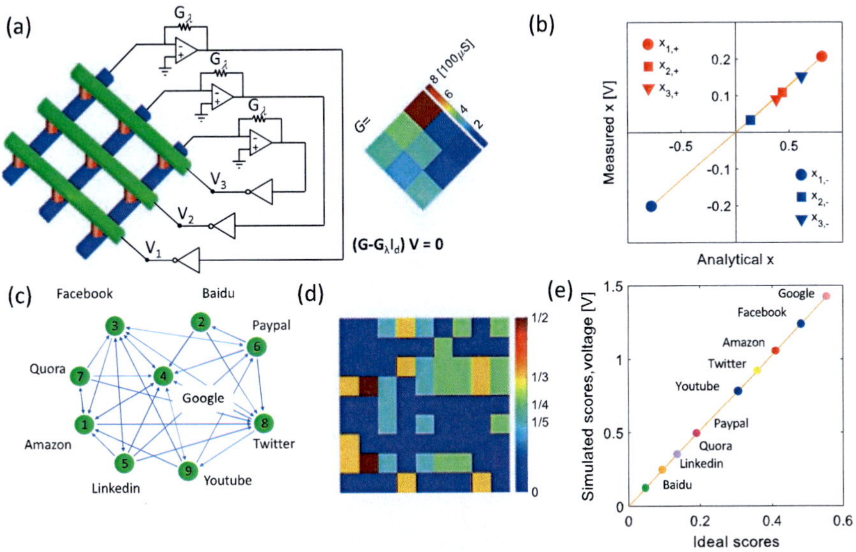

Fig. 11 In-memory eigenvectors calculation with IMVM. **a** IMVM circuit for eigenvector calculation, where no input is given and a conductance corresponding to the maximum eigenvalue G_λ is programmed in the TIA conductance. Inset shows a programmed measured matrix. **b** measured eigenvectors as function of the analytical calculation showing good agreement. **c** Graph of webpages used for Pagerank problem, where every circle is a webpage and the arrows represent citations. **d** Corresponding stochastic link matrix. **e** Simulated circuit result as function of the ideal scores showing good agreement. Reprinted with permission from [48] under Creative Commons License

4.1 In-Memory Eigenvector Calculation

By slightly modifying the topology of the analogue circuit in Fig. 10a, it is possible to calculate the principal eigenvector of the matrix programmed in the crosspoint [31, 48]. This is shown in Fig. 12a, where the matrix G is mapped in one crosspoint array and the principal eigenvalue λ of matrix G is mapped in the feedback resistor of the trans-impendence amplifier (TIA)s. A set of inverting OAs is then added in the feedback loop to compensate for the minus sign arising from the current–voltage conversion $V = -I/G_\lambda$ of the TIAs in Fig. 11a. The circuit is described by (3) with zero input current, thus leading to $(G - G_\lambda I)V = 0$, which corresponds to the non-trivial solution of the eigenvector problem for G. Figure 11b shows an experimental demonstration of the eigenvector circuit, showing the correlation plot of experimental components as a function of the ideal analytical values, for the eigenvectors corresponding to the maximum (principal) and the minimum eigenvalue [48].

The calculation of the principal eigenvector can be applied to relevant scientific computing tasks, such as the solution of the Schrödinger equation [48]. However,

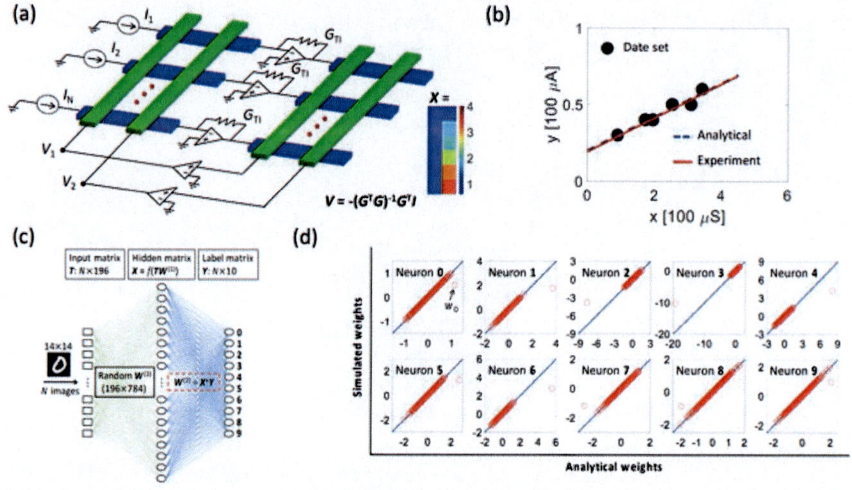

Fig. 12 In-memory regression calculation with IMVM. **a** IMVM circuit for Moore-Penrose pseudoinverse with a current I injected in a crosspoint array programmed with conductance G rows connected to a TIA whose output drives the rows of a second crosspoint array programmed with G^T. The column of the second crosspoint are connected to operational amplifiers whose outputs close the loop and are connected to the first crosspoint columns. The output voltage gives $V = -(G^T G)^{-1} G^T I$ which is the Moore Penrose pseudoinverse result. Inset shows a programmed linear regression problem. **b** measured fitting and analytical fitting of a programmed dataset showing good agreement. **c** ELM schematic for recognition of MNIST dataset with a random input layer and an output layer trained with logistic regression. **d** Circuit simulated weights as function of the analytical weights for the output layer showing a good agreement. Reprinted with permission from [107] under Creative Commons License

scientific computing requires high precision which is relatively difficult for in-memory computing due to imprecise RRAM programming. This problem can be circumvented by using IMVM for calculating a seed that is then refined with traditional computing technologies [106]. On the other hand, machine learning problems usually are less subject to noise and less sensitive to variations. For example, Pagerank [112], which is the algorithm that calculates webpage ranking on a search engine, requires the computation of the principal eigenvector of a link matrix corresponding to the adjacency between webpages as shown by the graph in Fig. 11c. Interestingly, the encoded matrix can be pre-processed to obtain a stochastic matrix (Fig. 11d) where the summation over all the columns is 1 and the principal eigenvalue is 1, thus making the IMVM circuit ideal for Pagerank calculation. Figure 11e shows a SPICE simulation of the Pagerank algorithm with IMVM compared with the analytical solution, highlighting the good agreement between the page scores [48]. The Pagerank problem particularly fits IMVM circuits also because the exact ranking is less important than the overall one, in fact users are usually interested in the first 10 webpages being displayed correctly in the Pagerank response, even if they are not listed in the correct order. For a more detailed assessment, the problem has been studied for a relatively large scale implementation with real conductance values programmed on HfO_x RRAM devices, showing a relatively low mismatch once a fine tuning of the conductance is performed [31]. The eigenvector calculation by IMVM has been shown to display a $O(1)$ time complexity, with an unprecedented speedup compared with other technologies [113].

4.2 *Pseudoinverse and Regression Accelerators*

Many machine learning problems can be written as the over-determined linear system $Xw = y$, where X is a rectangular $N \times M$ matrix with $N > M$ which encodes the explanatory variables, y is a $N \times 1$ known vector representing the dependent variables and w it the $M \times 1$ weight vector. Since this equation generally does not have a solution, its best approximation can be found by the linear regression, namely the least square error (LSE) algorithm that minimizes the Euclidean norm of the error, namely $\min\|Xw - y\|_2$. This minimization can be carried out by the pseudo-inverse, or Moore–Penrose inverse, namely matrix X^+ given by $X^+ = (X^T X)^{-1} X^T$, while the weights are given by $w = X^+ y$ [114]. Figure 12a shows an analogue IMVM circuit that can calculate the Moore–Penrose inverse matrix [107]. Two identical crosspoint arrays are used to map the matrix X in their conductance$G$. A vector of currents I representing the known term y is applied to the first crosspoint rows which are connected to the negative input of the TIAs with a feedback conductance G_{TI}. The first crosspoint columns are connected to the output terminals of a second stage of OAs with output voltage$V$. The first crosspoint execute the summation of input currents I and the MVM output GV, thus yielding an overall current $I + GV$ flowing into the TIAs. The latter develop a voltage across the second crosspoint array given by $-(I + VG)G_{TI}^{-1}$. The columns of the second crosspoint are connected to the positive inputs of the second stage of OAs thus must be equal to zero, which translates in the equation $(I + VG)G_{TI}^{-1}G^T = 0$ or equivalently:

$$V = -\left(G^T G\right)^{-1} G^T I \tag{4}$$

where the Moore–Penrose inverse matrix G^+ is clearly identified [107]. The inset of Fig. 12a shows the 2 × 6 matrix G which was mapped in a HfO_x RRAM crosspoint array representing the explanatory variables of a linear regression problem with the experimental solution that gives the best linear fit plotted in Fig. 12b and compared with the analytical calculation showing a good agreement [107].

This concept can be extended to the logistic regression which is a powerful classification algorithm. In fact, by properly writing in different columns of matrix G the coordinates of the data and injecting a relative current 1 or 0 corresponding to the class, it is possible to obtain the straight line that best separates the input data. This is a powerful tool in machine learning as it can be used for training the classification layer of a neural network. Figure 12c shows the conceptual schematic of a fully-connected, 2-layer neural network according to the Extreme Learning Machine (ELM) model [107], where all synaptic weights in the first layer are assumed to be random weights, while the synaptic weights in the output layer are trained by logistic regression. SPICE simulations of the IMVM circuit for training the output layer of the ELM model for classifying handwritten digits of the MNIST dataset [68] are shown in Fig. 12d and compared with the analytical solution. The results indicate a good agreement with the ideal solution, thus supporting the feasibility of IMVM circuits for training neural networks [107].

5 Conclusions

This chapter presents an overview of analogue in-memory computing concepts with RRAM devices. RRAM displays ideal properties for computing, including high density, analogue storage and the ability for 3D integration. MVM in the analogue domain is perhaps the most promising type of in-memory computing function which is made possible by a RRAM array, typically with 1T1R structure of the individual memory cell. Experiments and simulations show an unprecedented speed up of MVM for neural networks acceleration and CSP optimization, while IMVM displays strong advantages in terms of computational complexity and energy efficiency for algebraic and machine learning problems. At the same time, the RRAM technology and its operations should be optimized to fulfil all requirements of multibit operation, fast switching, controllable noise and long retention time necessary for enabling this technology in a relevant environment in the edge or cloud. In particular, the RRAM technology development and computing architecture research should proceed with strong synergy to fully take advantage of the energy and performance benefits of in-memory computing.

References

1. G.E. Moore, Cramming more components onto integrated circuits, Reprinted from Electronics, volume 38, number 8, April 19, 1965, pp.114 ff. IEEE Solid-State Circuits Soc. Newsl. **11**(3), 33–35 (2006). https://doi.org/10.1109/N-SSC.2006.4785860
2. S. Salahuddin, K. Ni, S. Datta, The era of hyper-scaling in electronics. Nat. Electron. **1**(8), 442–450 (2018). https://doi.org/10.1038/s41928-018-0117-x
3. D. Amodei, D. Hernandez, AI and compute, https://openai.com/blog/ai-and-compute/
4. J. von Neumann, First Draft of a Report on the EDVAC (1945). https://doi.org/10.5555/1102046
5. P.A. Merolla et al., A million spiking-neuron integrated circuit with a scalable communication network and interface. Science **345**(6197), 668–673 (2014). https://doi.org/10.1126/science.1254642
6. M.A. Zidan, J.P. Strachan, W.D. Lu, The future of electronics based on memristive systems. Nat. Electron. **1**(1), 22–29 (2018). https://doi.org/10.1038/s41928-017-0006-8
7. D. Ielmini, H.-S.P. Wong, In-memory computing with resistive switching devices. Nat. Electron. **1**(6), 333–343 (2018). https://doi.org/10.1038/s41928-018-0092-2
8. Z. Wang et al., Resistive switching materials for information processing. Nat. Rev. Mater. (2020). https://doi.org/10.1038/s41578-019-0159-3
9. L. Chua, Memristor-The missing circuit element. IEEE Trans. Circuit Theory **18**(5), 507–519 (1971). https://doi.org/10.1109/TCT.1971.1083337
10. D.B. Strukov, G.S. Snider, D.R. Stewart, R.S. Williams, The missing memristor found. Nature **453**(7191), 80–83 (2008). https://doi.org/10.1038/nature06932
11. D. Ielmini, Resistive switching memories based on metal oxides: mechanisms, reliability and scaling. Semicond. Sci. Technol. **31**(6), 063002 (2016). https://doi.org/10.1088/0268-1242/31/6/063002
12. T. Mikolajick et al., FeRAM technology for high density applications. Microelectron. Reliab. **41**(7), 947–950 (2001). https://doi.org/10.1016/S0026-2714(01)00049-X
13. J. Grollier, D. Querlioz, K.Y. Camsari, K. Everschor-Sitte, S. Fukami, M.D. Stiles, Neuromorphic spintronics. Nat. Electron. (2020). https://doi.org/10.1038/s41928-019-0360-9
14. Z. Sun, E. Ambrosi, A. Bricalli, D. Ielmini, Logic Computing with Stateful Neural Networks of Resistive Switches. Adv. Mater. **30**(38), 1802554 (2018). https://doi.org/10.1002/adma.201802554
15. G. Indiveri, B. Linares-Barranco, R. Legenstein, G. Deligeorgis, T. Prodromakis, Integration of nanoscale memristor synapses in neuromorphic computing architectures. Nanotechnology **24**(38), 384010 (2013). https://doi.org/10.1088/0957-4484/24/38/384010
16. C. Zamarreño-Ramos, L.A. Camuñas-Mesa, J.A. Pérez-Carrasco, T. Masquelier, T. Serrano-Gotarredona, B. Linares-Barranco, On Spike-Timing-Dependent-Plasticity, Memristive Devices, and Building a Self-Learning Visual Cortex. Front. Neurosci. **5** (2011). https://doi.org/10.3389/fnins.2011.00026
17. S.H. Jo, T. Chang, I. Ebong, B.B. Bhadviya, P. Mazumder, W. Lu, Nanoscale memristor device as synapse in neuromorphic systems. Nano Lett. **10**(4), 1297–1301 (2010). https://doi.org/10.1021/nl904092h
18. G. Pedretti et al., Memristive neural network for on-line learning and tracking with brain-inspired spike timing dependent plasticity. Sci. Rep. **7**(1), 5288 (2017). https://doi.org/10.1038/s41598-017-05480-0
19. W. Wang et al., Learning of spatiotemporal patterns in a spiking neural network with resistive switching synapses. Sci. Adv. **4**(9), eaat4752 (2018). https://doi.org/10.1126/sciadv.aat4752
20. R. Carboni, D. Ielmini, Stochastic memory devices for security and computing. Adv. Electron. Mater. **5**(9), 1900198 (2019). https://doi.org/10.1002/aelm.201900198
21. D. Ielmini, G. Pedretti, Device and Circuit architectures for in-memory computing. Adv. Intell. Syst. **2**(7), 2000040 (2020). https://doi.org/10.1002/aisy.202000040
22. H.-S.P. Wong et al., Metal–Oxide RRAM. Proc. IEEE **100**(6), 1951–1970 (2012). https://doi.org/10.1109/JPROC.2012.2190369

23. S. Raoux, W. Wełnic, D. Ielmini, Phase change materials and their application to nonvolatile memories. Chem. Rev. **110**(1), 240–267 (2010). https://doi.org/10.1021/cr900040x
24. G.W. Burr et al., Phase change memory technology. J. Vac. Sci. Technol., B: Nanotechnol. Microelectron.: Mater. Process. Meas. Phenom. **28**(2), 223–262 (2010). https://doi.org/10.1116/1.3301579
25. T. S. Boscke, J. Muller, D. Brauhaus, U. Schroder, U. Bottger, Ferroelectricity in hafnium oxide: CMOS compatible ferroelectric field effect transistors, in *2011 International Electron Devices Meeting* (Washington, DC, USA, 2011), pp. 24.5.1–24.5.4. https://doi.org/10.1109/IEDM.2011.6131606
26. H. Mulaosmanovic et al., Novel ferroelectric FET based synapse for neuromorphic systems, in *2017 Symposium on VLSI Technology* (Kyoto, Japan, 2017), pp. T176–T177. https://doi.org/10.23919/VLSIT.2017.7998165
27. C. Chappert, A. Fert, F.N. Van Dau, The emergence of spin electronics in data storage. Nat. Mater. **6**, 813–823 (2007)
28. B. Govoreanu et al., 10x10nm^2 Hf/HfOx crossbar resistive RAM with excellent performance, reliability and low-energy operation, in *2011 International Electron Devices Meeting* (Washington, DC, USA, 2011), pp. 31.6.1–31.6.4. https://doi.org/10.1109/IEDM.2011.6131652
29. A.C. Torrezan, J.P. Strachan, G. Medeiros-Ribeiro, R.S. Williams, Sub-nanosecond switching of a tantalum oxide memristor. Nanotechnology **22**(48), 485203 (2011). https://doi.org/10.1088/0957-4484/22/48/485203
30. S. Yu, H.-Y. Chen, B. Gao, J. Kang, H.-S.P. Wong, HfO_x-based vertical resistive switching random access memory suitable for bit-cost-effective three-dimensional cross-point architecture. ACS Nano **7**(3), 2320–2325 (2013). https://doi.org/10.1021/nn305510u
31. Z. Sun, E. Ambrosi, G. Pedretti, A. Bricalli, D. Ielmini, In-Memory PageRank Accelerator With a Cross-Point Array of Resistive Memories. IEEE Trans. Electron Devices **67**(4), 1466–1470 (2020). https://doi.org/10.1109/TED.2020.2966908
32. J.J. Yang, D.B. Strukov, D.R. Stewart, Memristive devices for computing. Nature Nanotech. **8**(1), 13–24 (2013). https://doi.org/10.1038/nnano.2012.240
33. S. N. Truong, K.-S. Min, New memristor-based crossbar array architecture with 50-% area reduction and 48-% power saving for matrix-vector multiplication of analog neuromorphic computing. JSTS: J. Semicond. Technol. Sci. **14**(3), 356–363 (2014). https://doi.org/10.5573/JSTS.2014.14.3.356
34. M. Hu et al., Memristor-based analog computation and neural network classification with a dot product engine. Adv. Mater. **30**(9), 1705914 (2018). https://doi.org/10.1002/adma.201705914
35. M.-C. Hsieh et al., Ultra high density 3D via RRAM in pure 28nm CMOS process, in *2013 IEEE International Electron Devices Meeting* (Washington, DC, USA, 2013), pp. 10.3.1–10.3.4. https://doi.org/10.1109/IEDM.2013.6724600
36. E. Linn, R. Rosezin, C. Kügeler, R. Waser, Complementary resistive switches for passive nanocrossbar memories. Nat. Mater. **9**(5), 403–406 (2010). https://doi.org/10.1038/nmat2748
37. D. Ielmini, Y. Zhang, Physics-based analytical model of chalcogenide-based memories for array simulation, in *2006 International Electron Devices Meeting* (San Francisco, CA, USA, 2006), pp. 1–4. https://doi.org/10.1109/IEDM.2006.346795
38. L. Gao, P.-Y. Chen, R. Liu, S. Yu, Physical unclonable function exploiting sneak paths in resistive cross-point array. IEEE Trans. Electron Devices **63**(8), 3109–3115 (2016). https://doi.org/10.1109/TED.2016.2578720
39. F. Li, X. Yang, A.T. Meeks, J.T. Shearer, K.Y. Le, Evaluation of SiO_2 antifuse in a 3D-OTP memory. IEEE Trans. Device Mater. Relib. **4**(3), 416–421 (2004). https://doi.org/10.1109/TDMR.2004.837118
40. Tz-Yi Liu et al., A 130.7mm^2 2-layer 32Gb ReRAM memory device in 24nm technology, in *2013 IEEE International Solid-State Circuits Conference Digest of Technical Papers* (San Francisco, CA, 2013), pp. 210–211. https://doi.org/10.1109/ISSCC.2013.6487703

41. G.W. Burr et al., Access devices for 3D crosspoint memory. J. Vac. Sci. Technol. B, Nanotechnolo. Microelectronics: Mater. Process. Meas. Phenom. **32**(4), 040802 (2014). https://doi.org/10.1116/1.4889999
42. D. Ielmini, Modeling the universal set/reset characteristics of bipolar rram by field- and temperature-driven filament growth. IEEE Trans. Electron Devices **58**(12), 4309–4317 (2011). https://doi.org/10.1109/TED.2011.2167513
43. V. Milo et al., Multilevel HfO_2-based RRAM devices for low-power neuromorphic networks. APL Mater. **7**(8), 081120 (2019). https://doi.org/10.1063/1.5108650
44. C. Li et al., Analogue signal and image processing with large memristor crossbars. Nat Electron **1**(1), 52–59 (2018). https://doi.org/10.1038/s41928-017-0002-z
45. A. Shafiee et al., ISAAC: a convolutional neural network accelerator with in-situ analog arithmetic in crossbars, in *2016 ACM/IEEE 43rd Annual International Symposium on Computer Architecture (ISCA)* (Seoul, South Korea, 2016), pp. 14–26. https://doi.org/10.1109/ISCA.2016.12
46. S. Balatti, S. Ambrogio, D.C. Gilmer, D. Ielmini, Set variability and failure induced by complementary switching in bipolar RRAM. IEEE Electron Device Lett. **34**(7), 861–863 (2013). https://doi.org/10.1109/LED.2013.2261451
47. S. Ambrogio, S. Balatti, A. Cubeta, A. Calderoni, N. Ramaswamy, D. Ielmini, Statistical fluctuations in HfO_x resistive-switching memory: Part I-set/reset variability. IEEE Trans. Electron Devices **61**(8), 2912–2919 (2014). https://doi.org/10.1109/TED.2014.2330200
48. Z. Sun, G. Pedretti, E. Ambrosi, A. Bricalli, W. Wang, D. Ielmini, Solving matrix equations in one step with cross-point resistive arrays. Proc Natl Acad Sci USA **116**(10), 4123–4128 (2019). https://doi.org/10.1073/pnas.1815682116
49. Y.-H. Lin et al., Performance impacts of analog ReRAM non-ideality on neuromorphic computing. IEEE Trans. Electron Devices **66**(3), 1289–1295 (2019). https://doi.org/10.1109/TED.2019.2894273
50. S. Balatti et al., Voltage-controlled cycling endurance of HfO_x-based resistive-switching memory. IEEE Trans. Electron Devices **62**(10), 3365–3372 (2015). https://doi.org/10.1109/TED.2015.2463104
51. S. Ambrogio, S. Balatti, V. McCaffrey, D.C. Wang, D. Ielmini, Noise-induced resistance broadening in resistive switching memory—Part II: array statistics. IEEE Trans. Electron Devices **62**(11), 3812–3819 (2015). https://doi.org/10.1109/TED.2015.2477135
52. C. Li et al., Efficient and self-adaptive in-situ learning in multilayer memristor neural networks. Nat. Commun. **9**(1), 2385 (2018). https://doi.org/10.1038/s41467-018-04484-2
53. P. Yao et al., Face classification using electronic synapses. Nat. Commun. **8**(1), 15199 (2017). https://doi.org/10.1038/ncomms15199
54. Z. Wang et al., Reinforcement learning with analogue memristor arrays. Nat. Electron. **2**(3), 115–124 (2019). https://doi.org/10.1038/s41928-019-0221-6
55. Z. Wang et al., In situ training of feed-forward and recurrent convolutional memristor networks. Nat. Mach. Intell. **1**(9), 434–442 (2019). https://doi.org/10.1038/s42256-019-0089-1
56. P.M. Sheridan, F. Cai, C. Du, W. Ma, Z. Zhang, W.D. Lu, Sparse coding with memristor networks. Nat. Nanotech. **12**(8), 784–789 (2017). https://doi.org/10.1038/nnano.2017.83
57. M.R. Mahmoodi, M. Prezioso, D.B. Strukov, Versatile stochastic dot product circuits based on nonvolatile memories for high performance neurocomputing and neurooptimization. Nat. Commun. **10**(1), 5113 (2019). https://doi.org/10.1038/s41467-019-13103-7
58. J.H. Shin, Y.J. Jeong, M.A. Zidan, Q. Wang, W.D. Lu, Hardware Acceleration of simulated annealing of spin glass by RRAM crossbar array, in *2018 IEEE International Electron Devices Meeting (IEDM)* (San Francisco, CA, 2018), pp. 3.3.1–3.3.4. https://doi.org/10.1109/IEDM.2018.8614698
59. F. Cai et al., Power-efficient combinatorial optimization using intrinsic noise in memristor Hopfield neural networks. Nat. Electron. (2020). https://doi.org/10.1038/s41928-020-0436-6
60. K. Yang, Q. Duan, Y. Wang, T. Zhang, Y. Yang, R. Huang, Transiently chaotic simulated annealing based on intrinsic nonlinearity of memristors for efficient solution of optimization problems. Sci. Adv. **6**(33), pp. eaba9901 (2020). https://doi.org/10.1126/sciadv.aba9901

61. M.R. Mahmoodi, D.B. Strukov, O. Kavehei, Experimental demonstrations of security primitives with nonvolatile memories. IEEE Trans. Electron Devices **66**(12), 5050–5059 (2019). https://doi.org/10.1109/TED.2019.2948950
62. H. Nili et al., Hardware-intrinsic security primitives enabled by analogue state and nonlinear conductance variations in integrated memristors. Nat. Electron. **1**(3), 197–202 (2018). https://doi.org/10.1038/s41928-018-0039-7
63. M.A. Zidan et al., A general memristor-based partial differential equation solver. Nat. Electron. **1**(7), 411–420 (2018). https://doi.org/10.1038/s41928-018-0100-6
64. P. Yao et al., Fully hardware-implemented memristor convolutional neural network. Nature **577**(7792), 641–646 (2020). https://doi.org/10.1038/s41586-020-1942-4
65. F. Cai et al., A fully integrated reprogrammable memristor–CMOS system for efficient multiply–accumulate operations. Nat. Electron. **2**(7), 290–299 (2019). https://doi.org/10.1038/s41928-019-0270-x
66. C. Li et al., CMOS-integrated nanoscale memristive crossbars for CNN and optimization acceleration, in *2020 IEEE International Memory Workshop (IMW)* (Dresden, Germany, 2020), pp. 1–4. https://doi.org/10.1109/IMW48823.2020.9108112
67. S. Yin, S. Yu, High-throughput in-memory computing for binary deep neural networks with monolithically integrated RRAM and 90-nm CMOS. IEEE Trans. Electron Devices **67**(10), 8 (2020)
68. Y. LeCun, Y. Bengio, G. Hinton, Deep learning. Nature **521**(7553), 436–444 (2015). https://doi.org/10.1038/nature14539
69. P. Chi et al., PRIME: a novel processing-in-memory architecture for neural network computation in ReRAM-Based main memory, in *2016 ACM/IEEE 43rd Annual International Symposium on Computer Architecture (ISCA)* (Seoul, South Korea, 2016), pp. 27–39. https://doi.org/10.1109/ISCA.2016.13
70. T. Gokmen, Y. Vlasov, Acceleration of Deep Neural Network Training with Resistive Cross-Point Devices: Design Considerations. Front. Neurosci. **10** (2016). https://doi.org/10.3389/fnins.2016.00333
71. S. Yu, Neuro-inspired computing with emerging nonvolatile memorys. Proc. IEEE **106**(2), 260–285 (2018). https://doi.org/10.1109/JPROC.2018.2790840
72. S. Ambrogio et al., Equivalent-accuracy accelerated neural-network training using analogue memory. Nature **558**(7708), 60–67 (2018). https://doi.org/10.1038/s41586-018-0180-5
73. G.W. Burr et al., Experimental demonstration and tolerancing of a large-scale neural network (165 000 Synapses) using phase-change memory as the synaptic weight element. IEEE Trans. Electron Devices **62**(11), 3498–3507 (2015). https://doi.org/10.1109/TED.2015.2439635
74. H. Kim et al., Zero-shifting technique for deep neural network training on resistive cross-point arrays. arXiv:1907.10228 [cs.ET] (2019)
75. S. Kim et al., Metal-oxide based, CMOS-compatible ECRAM for Deep Learning Accelerator, in *2019 IEEE International Electron Devices Meeting (IEDM)* (San Francisco, CA, USA, 2019), pp. 35.7.1–35.7.4. https://doi.org/10.1109/IEDM19573.2019.8993463
76. C.-C. Chang et al., Challenges and opportunities toward online training acceleration using rram-based hardware neural network," in *2017 IEEE International Electron Devices Meeting (IEDM)* (San Francisco, CA, USA, 2017), pp. 11.6.1–11.6.4
77. Z. Zhou et al., A new hardware implementation approach of BNNs based on nonlinear 2T2R synaptic cell, in *2018 IEEE International Electron Devices Meeting (IEDM)* (San Francisco, CA, 2018), pp. 20.7.1–20.7.4. https://doi.org/10.1109/IEDM.2018.8614642
78. D. Garbin et al., HfO_2-based OxRAM devices as synapses for convolutional neural networks. IEEE Trans. Electron Devices **62**(8), 2494–2501 (2015). https://doi.org/10.1109/TED.2015.2440102
79. I. Boybat et al., Neuromorphic computing with multi-memristive synapses. Nat. Commun. **9**(1), 2514 (2018). https://doi.org/10.1038/s41467-018-04933-y
80. V. Milo et al., A 4-Transistors/1-resistor hybrid synapse based on resistive switching memory (RRAM) capable of spike-rate-dependent plasticity (SRDP). IEEE Trans. VLSI Syst. **26**(12), 2806–2815 (2018). https://doi.org/10.1109/TVLSI.2018.2818978

81. Z. Wang et al., Toward a generalized Bienenstock-Cooper-Munro rule for spatiotemporal learning via triplet-STDP in memristive devices. Nat. Commun. **11**(1), 1510 (2020). https://doi.org/10.1038/s41467-020-15158-3
82. Z. Wang et al., Fully memristive neural networks for pattern classification with unsupervised learning. Nat. Electron. **1**(2), 137–145 (2018). https://doi.org/10.1038/s41928-018-0023-2
83. S. Liu, Y. Wang, M. Fardad, P.K. Varshney, A memristor-based optimization framework for artificial intelligence applications. IEEE Circuits Syst. Mag. **18**(1), 29–44 (2018). https://doi.org/10.1109/MCAS.2017.2785421
84. J. Hopfield, D. Tank, Computing with neural circuits: a model. Science **233**(4764), 625–633 (1986). https://doi.org/10.1126/science.3755256
85. J.J. Hopfield, Neurons with graded response have collective computational properties like those of two-state neurons. Proc. Natl. Acad. Sci. **81**(10), 3088–3092 (1984). https://doi.org/10.1073/pnas.81.10.3088
86. S. B. Eryilmaz et al., Brain-like associative learning using a nanoscale non-volatile phase change synaptic device array. Front. Neurosci. **8** (2014). https://doi.org/10.3389/fnins.2014.00205
87. V. Milo, D. Ielmini, E. Chicca, Attractor networks and associative memories with STDP learning in RRAM synapses, in *2017 IEEE International Electron Devices Meeting (IEDM)* (San Francisco, CA, USA, 2017), pp. 11.2.1–11.2.4. https://doi.org/10.1109/IEDM.2017.8268369
88. G. Pedretti et al., A spiking recurrent neural network with phase change memory synapses for decision making, in *2020 IEEE International Symposium on Circuits and Systems (ISCAS)* (Sevilla, 2020), pp. 1–5. https://doi.org/10.1109/ISCAS45731.2020.9180513
89. J.J. Hopfield, D.W. Tank, Neural computation of decisions in optimization problems. Biol. Cybern. **52**, 141–152 (1985)
90. A. Lucas, Ising formulations of many NP problems. Front. Phys. **2** (2014). https://doi.org/10.3389/fphy.2014.00005
91. S. Kirkpatrick, C.D. Gelatt, M.P. Vecchi, Optimization by Simulated Annealing. Science **220**(4598), 671–680 (1983)
92. G.A. Fonseca Guerra, S.B. Furber, Using stochastic spiking neural networks on spinnaker to solve constraint satisfaction problems, Front. Neurosci. **11**, 714 (2017). https://doi.org/10.3389/fnins.2017.00714
93. H. Mostafa, L.K. Müller, G. Indiveri, An event-based architecture for solving constraint satisfaction problems. Nat Commun **6**(1), 8941 (2015). https://doi.org/10.1038/ncomms9941
94. T. Takemoto, M. Hayashi, C. Yoshimura, M. Yamaoka, 2.6 A 2 $\times$30k-Spin multichip scalable annealing processor based on a processing-in-memory approach for solving large-scale combinatorial optimization problems, in *2019 IEEE International Solid- State Circuits Conference—(ISSCC)* (San Francisco, CA, USA, 2019), pp. 52–54. https://doi.org/10.1109/ISSCC.2019.8662517
95. F.L. Traversa, C. Ramella, F. Bonani, M. Di Ventra, Memcomputing *NP*—complete problems in polynomial time using polynomial resources and collective states. Sci. Adv. **1**(6), e1500031 (2015). https://doi.org/10.1126/sciadv.1500031
96. V. S. Denchev et al., What is the computational value of finite-range tunneling?. Phys. Rev. X, **6**(3), 031015 (2016). https://doi.org/10.1103/PhysRevX.6.031015
97. S. Boixo et al., Evidence for quantum annealing with more than one hundred qubits. Nat. Phys. **10**(3), 218–224 (2014). https://doi.org/10.1038/nphys2900
98. P. L. McMahon et al., A fully programmable 100-spin coherent Ising machine with all-to-all connections, pp. 5
99. S. Kumar, J.P. Strachan, R.S. Williams, Chaotic dynamics in nanoscale NbO2 Mott memristors for analogue computing. Nature **548**(7667), 318–321 (2017). https://doi.org/10.1038/nature23307
100. G. Pedretti et al., A spiking recurrent neural network with phase-change memory neurons and synapses for the accelerated solution of constraint satisfaction problems. IEEE. J Explor. Solid-State Comput. Devices Circuits **6**(1), 89–97 (2020). https://doi.org/10.1109/JXCDC.2020.2992691

101. S. Kumar, R.S. Williams, Z. Wang, Third-order nanocircuit elements for neuromorphic engineering. Nature **585**(7826), 518–523 (2020). https://doi.org/10.1038/s41586-020-2735-5
102. M.R. Mahmoodi et al., An analog neuro-optimizer with adaptable annealing based on 64x64 0t1r crossbar circuit, in *2019 IEEE International Electron Devices Meeting (IEDM)* (San Francisco, CA, 2019), pp. 14.7.1–14.7.4. https://doi.org/10.1109/IEDM19573.2019.8993442
103. S. Ambrogio, S. Balatti, A. Cubeta, A. Calderoni, N. Ramaswamy, D. Ielmini, Statistical fluctuations in HfO_x resistive-switching memory: Part II—random telegraph noise. IEEE Trans. Electron Devices **61**(8), 2920–2927 (2014). https://doi.org/10.1109/TED.2014.2330202
104. S. Ambrogio, S. Balatti, V. McCaffrey, D.C. Wang, D. Ielmini, Noise-induced resistance broadening in resistive switching memory—Part I: intrinsic cell behavior. IEEE Trans. Electron Devices **62**(11), 3805–3811 (2015). https://doi.org/10.1109/TED.2015.2475598
105. L. Chen, K. Aihara, Chaotic simulated annealing by a neural network model with transient chaos. Neural Netw. **8**(6), 915–930 (1995). https://doi.org/10.1016/0893-6080(95)00033-V
106. M. Le Gallo et al., Mixed-precision in-memory computing. Nat Electron **1**(4), 246–253 (2018). https://doi.org/10.1038/s41928-018-0054-8
107. Z. Sun, G. Pedretti, A. Bricalli, D. Ielmini, One-step regression and classification with cross-point resistive memory arrays, Sci. Adv. **6**(5), eaay2378 (2020). https://doi.org/10.1126/sciadv.aay2378
108. Z. Sun, G. Pedretti, P. Mannocci, E. Ambrosi, A. Bricalli, D. Ielmini, Time complexity of in-memory solution of linear systems. IEEE Trans. Electron Devices **67**(7), 2945–2951 (2020). https://doi.org/10.1109/TED.2020.2992435
109. Z. Sun, G. Pedretti, D. Ielmini, Fast solution of linear systems with analog resistive switching memory (RRAM), in *2019 IEEE International Conference on Rebooting Computing (ICRC)* (San Mateo, CA, USA, 2019), pp. 1–5. https://doi.org/10.1109/ICRC.2019.8914709
110. J.R. Shewchuk, An Introduction to the Conjugate Gradient Method without the Agonizing Pain, School of Computer Science, Carnegie Mellon University, Pittsburgh, CMU-CS-94-125, (1994)
111. A. W. Harrow, A. Hassidim, S. Lloyd, Quantum algorithm for linear systems of equations. Phys. Rev. Lett.**103**(15), 150502 (2009). https://doi.org/10.1103/PhysRevLett.103.150502.
112. K. Bryan, T. Leise, The $25,000,000,000 eigenvector: the linear algebra behind google. SIAM Rev. **48**(3), 569–581 (2006). https://doi.org/10.1137/050623280
113. Z. Sun, G. Pedretti, E. Ambrosi, A. Bricalli, D. Ielmini, In-memory eigenvector computation in time O (1). Adv. Intell. Syst. 2000042 (2020). https://doi.org/10.1002/aisy.202000042
114. R. Penrose, A generalized inverse for matrices. Math. Proc. Camb. Phil. Soc. **51**(3), 406–413 (1955). https://doi.org/10.1017/S0305004100030401

Introduction to 3D NAND Flash Memories

Rino Micheloni, Luca Crippa, and Cristian Zambelli

Abstract Nowadays, NAND Flash memories are in everybody's hands, as they are the storage media used inside smartphones and tablets. At the same time NAND Flash memories, in the form of Solid State Drives (SSDs), have enabled a new generation of computers without Hard Disk Drives (HDDs), and they are also one of the key components of the modern cloud infrastructures. To win all these new applications, NAND Flash had to continuously decrease its cost per bit. Shrinking lithography has been the solution for many generations of planar NANDs, but this approach ran out of steam in the sub-20nm range due to a plethora of parasitic effects within the memory array. As such, both the industry and the academia have worked towards a different approach for many years, resulting in a tri-dimensional (3D) architecture, whose first product reached the market in 2016. In this Chapter we present the basics of 3D NAND Flash memories and the related integration challenges. There are two main variants of Flash technologies used inside 3D arrays, namely, Floating Gate (FG) and Charge Trap (CT), which are both described in this Chapter with the aid of several bird's-eye views. Finally, 3D scaling trends are discussed.

Nowadays, Solid State Drives consume an enormous amount of NAND Flash memories [1] causing a restless pressure on increasing the number of stored bits per mm^2.

This chapter is an authorized partial reprint of Micheloni R., Aritome S., Crippa L. (2018) 3D NAND Flash Memories. In: Micheloni R., Marelli A., Eshghi K. (eds) Inside Solid State Drives (SSDs). Springer Series in Advanced Microelectronics, vol 37. Springer, Singapore. https://doi.org/10.1007/978-981-13-0599-3_5.

R. Micheloni (✉) · C. Zambelli
Dipartimento di Ingegneria, Università degli Studi di Ferrara, Via G. Saragat 1, 44122 Ferrara, Italy
e-mail: rino.micheloni@ieee.org

C. Zambelli
e-mail: cristian.zambelli@unife.it

L. Crippa
Manzoni 66, 20874 Busnago (MB), Italy
e-mail: luca.crippa@ieee.org

R. Micheloni and C. Zambelli (eds.), *Machine Learning and Non-volatile Memories*,
https://doi.org/10.1007/978-3-031-03841-9_5

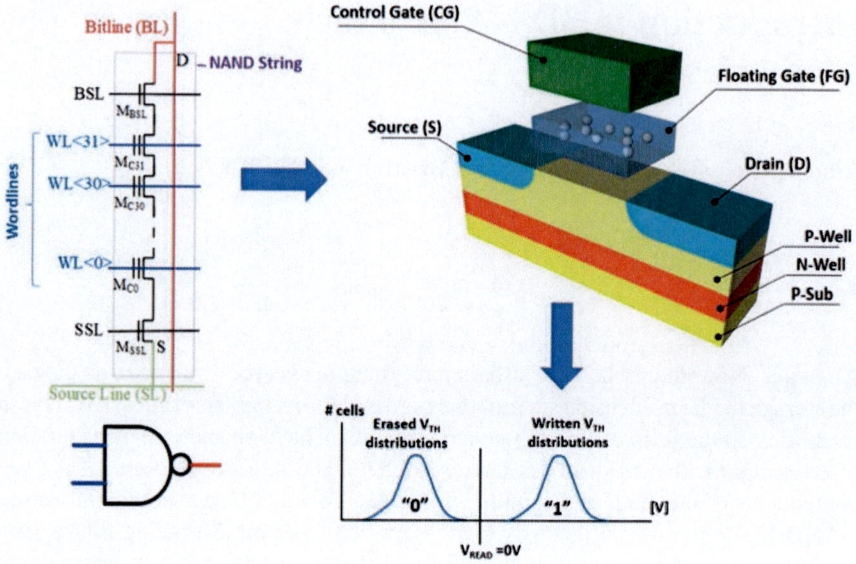

Fig. 1 NAND string

Planar memory cells (Fig. 1) have been scaled for decades by improving process technology, circuit design, programming algorithms [2], and lithography.

Unfortunately, when approaching a minimum feature size of 1x-nm, more challenges pop up: doping concentration in the channel region becomes difficult to control [3], RTN [4] and electron injection statistics [5] widen threshold distributions, thus causing a significant hit to both endurance and retention. Furthermore, by reducing the distance between memory cells, the intra-wordline electric field becomes higher, pushing the bit error rate to an even higher level.

3D arrays can definitely be considered as a breakthrough for fueling a further increase of the bit density. Identifying the right way for going 3D was not so easy though.

Historically, Flash memory manufacturers have leveraged lithography to shrink the 2-dimensional (2D) memory cell [6].

However, with 3D architectures, the "simple" reduction of the minimum feature size is running out of steam [7]: a higher number of stacked cells is the only hope for dramatically reducing the real estate of a stored bit.

3D arrays can leverage either *Floating Gate* (FG) or *Charge Trapping* (CT) technologies [8]. As a matter of fact, the vast majority of 3D architectures published to date are built with CT cells, mainly because of the simpler fabrication process. Nevertheless, Floating Gate is still around and there are commercial products who managed to integrate FG into a 3D array.

1 3D Charge Trap NAND Flash Memories

3D arrays can be efficiently built by vertically rotating the planar NAND Flash string as displayed in Fig. 2. The solution of choice is a conduction channel completely surrounded by the gate [9]: indeed, the curvature effect helps increasing the electric field E_t across the tunnel oxide, and reduces the electric field E_b across the blocking oxide [10, 11], and this has a positive impact on oxide reliability and overall power consumption.

Vertical channel arrays have been historically driven by architectures known as BiCS, which stands for *Bit Cost Scalable* [12, 13] and P-BiCS, acronym for *Pipe-Shaped BiCS* [14–16], which are both leveraging CT cells [17]. Let's get started with BiCS, which is sketched in Figs. 3 and 4 [13]. There is a stack of *Control Gates* (CGs), the lowest being the one of the *Source Line Selector* (SLS). The whole vertical stack is punched through and the resulting holes are filled with poly-silicon; each filled hole (a.k.a. pillar) forms a series of memory cells vertically connected in a NAND fashion. *Bit Line Selectors* (BLS's) and *Bitlines* (BLs) are formed at the top of the structure [18].

The poly-silicon body of memory cells is not doped or lightly doped [10, 11]; indeed, considering the bad aspect ratio of the vertical polysilicon plug, p-n junctions cannot be easily realized by either diffusion or implantation in a trench structure. As usual, a select transistor (BLS) is used to connect each NAND string to a bitline;

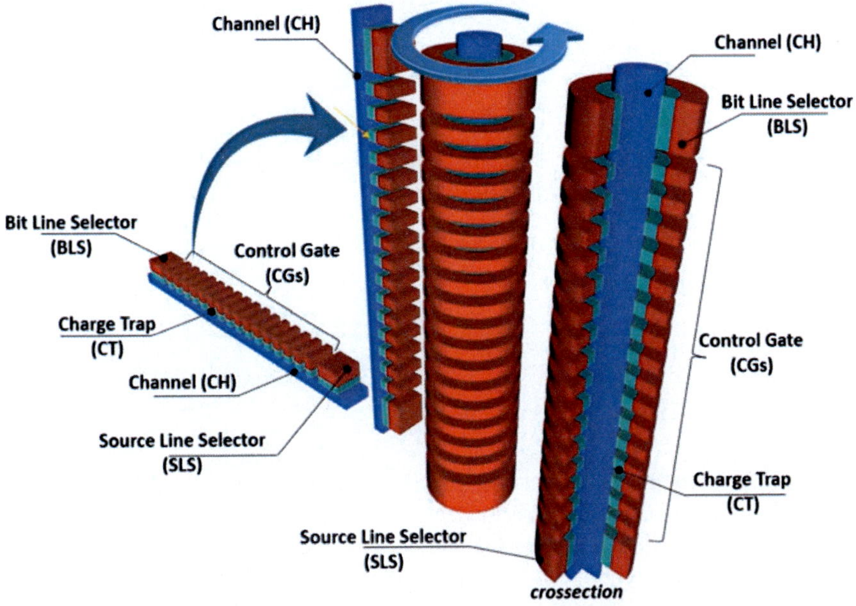

Fig. 2 The NAND flash string goes vertical

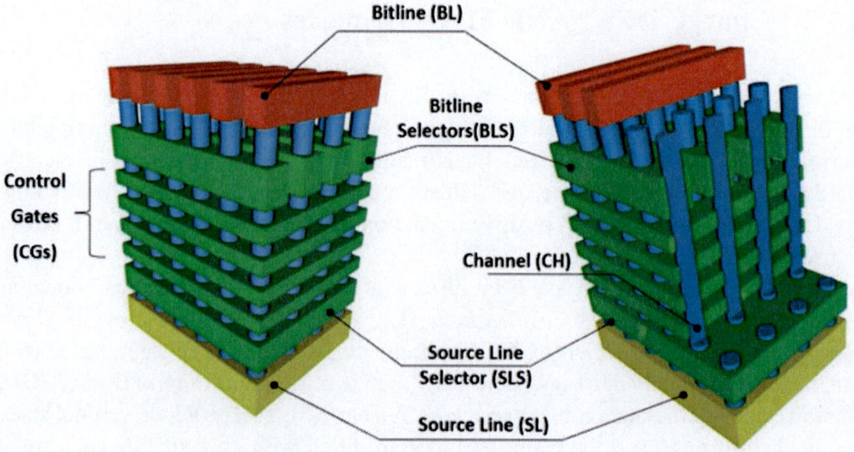

Fig. 3 BiCS architecture. Adapted with permission from [19]. ©2017 IEEE

Fig. 4 Equivalent circuit of a BiCS array

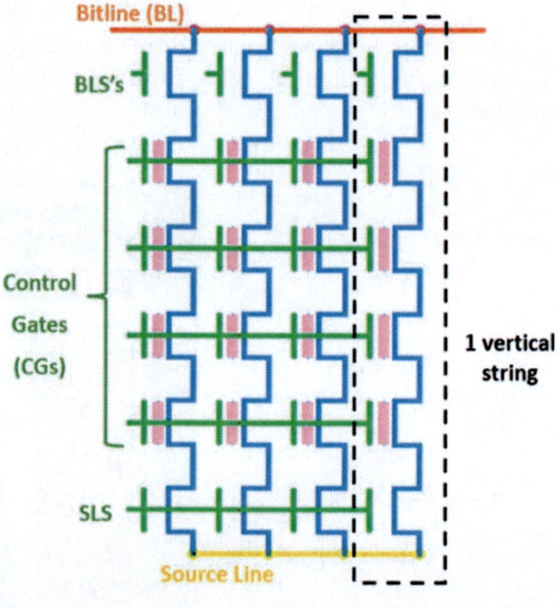

there is also another select transistor (SLS), which connects the other side of the string to the common source diffusion.

It is important to highlight that the number of critical and expensive lithography steps does not depend on the number of control gate plates because the whole 3D stack is drilled at one [20, 21].

As sketched in Fig. 5, vertical transistor have polysilicon body and this fact turned out to be one of the critical cornerstone of the 3D foundation. From a manufacturing

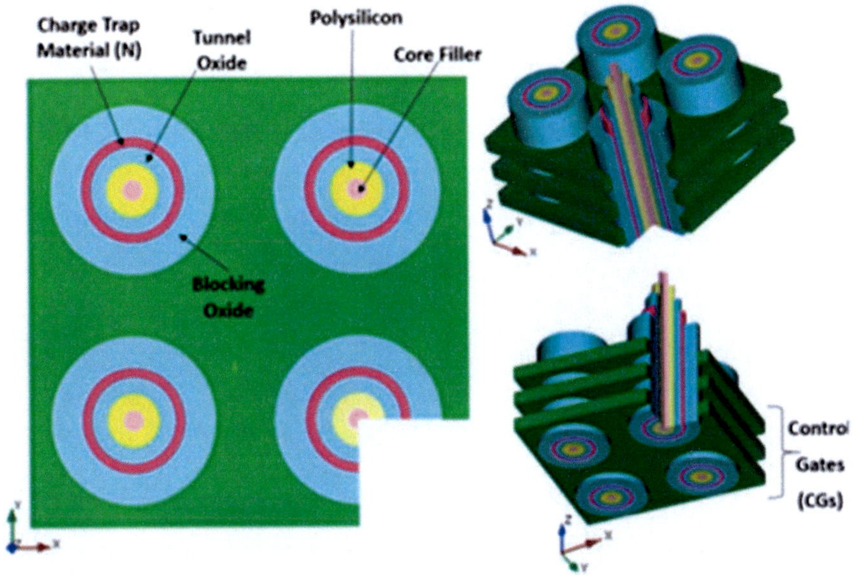

Fig. 5 BiCS memory cells

perspective, the density of the traps at the grain boundary is very difficult to control, with such a vertical shape: the bad thing is that this poor control induces significant fluctuations of the characteristics of vertical transistors.

The recipe for fixing the trap density fluctuation problem is to manufacture a polysilicon body much thinner than the depletion width. In other words, by shrinking the polysilicon volume, the total number of traps goes down (Fig. 6). This particular structure is usually referred to as *Macaroni Body* [13]. A *filler layer* (i.e. a dielectric film) is used in the central part of the macaroni structure, essentially because it makes the manufacturing process easier.

The fabrication sequence of the BiCS array [22] starts from building the layers for control gates and selectors. Then, BLS stripes are defined. After forming pillars, bitlines are laid out by using a metal layer.

Control gate edges are extended to form a ladder to connect to the fan-out region, as sketched in Fig. 7 [12, 13, 22, 23]. Actually, there are 2 ladders: one of the 2 can't be used because it is masked by the metals biasing the bitline selectors.

Over time BiCS became P-BiCS, mainly to improve the Source Line resistance [14, 15]. In a nutshell, two vertical NAND strings are shorted together at the bottom of the 3D structure: in this way, they form a single NAND string and the 2 edges are connected to the bitline and to the Source Line, respectively (Fig. 8). Thanks to its U-shape, P-BiCS has few advantages over BiCS:

- retention is better because manufacturing creates less damages in the tunnel oxide;
- being at the top, the Source Line can be connected to a metal mesh, thus lowering its parasitic resistance;

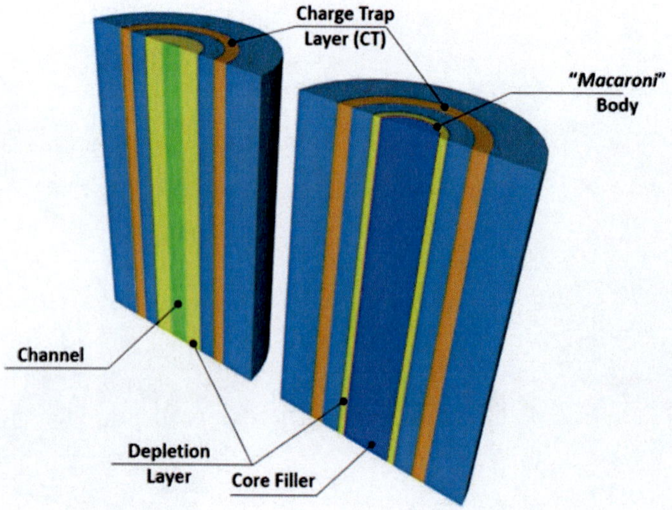

Fig. 6 A vertical transistor (right) modified with *Macaroni* body (left)

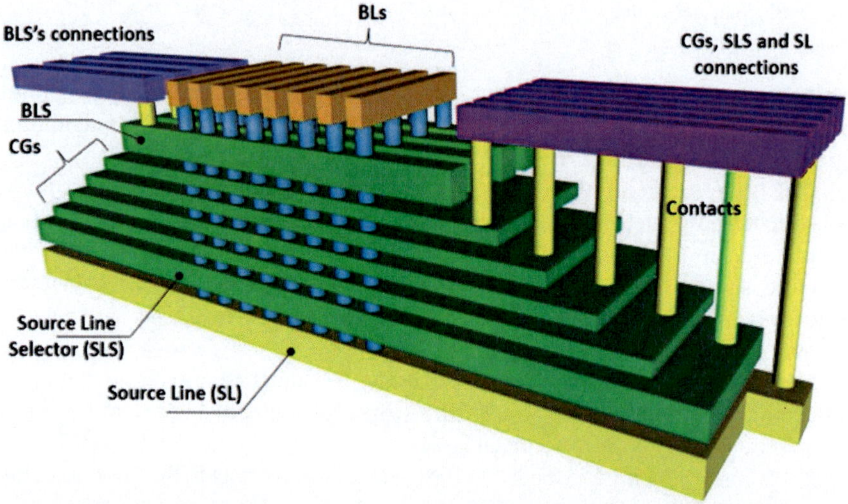

Fig. 7 Fan-out of the BiCS array. Adapted with permission from [19]. ©2017 IEEE

- Source Line and bitline selectors are at the same height of the stack and, therefore, they can be equally optimized and controlled, thus obtaining a better string functionality.

One of the biggest drawbacks of P-BiCS is the fact that at the same height of the stack there are two different control gates which, of course, can't be biased together; therefore, the two layers can't be simply shorted together. As a result, compared to

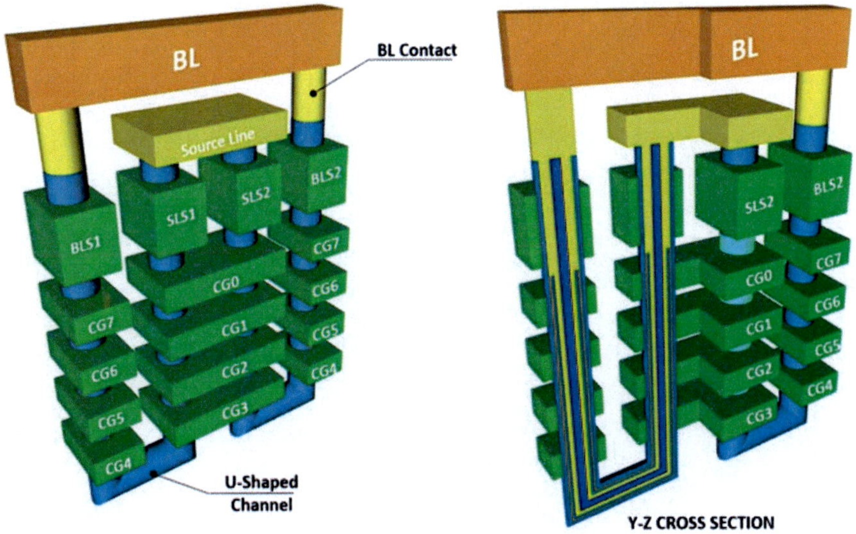

Fig. 8 P-BICS NAND strings

BiCS, a totally different and more complex fan-out is required [16], as displayed in Fig. 9: basically, a fork-shaped gate is adopted, such that each branch acts on two NAND pages.

A major advantage is the easier connection of the source line [14] through the "Top Level Source Line" of Fig. 10. This additional metal mesh guarantees a much better noise immunity for circuits.

Besides BiCS and P-BiCS, many other approaches were tried, including VRAT (*Vertical Recess Array Transistor*) [24], Z-VRAT (*Zigzag* VRAT) [24], and VSAT (*Vertical Stacked Array Transistor*) [25], and 3D-VG (*Vertical Gate*) NAND [26] which is a unique architecture where the channel runs along the horizontal direction.

TCAT (*Terabit Cell Array Transistor*) was disclosed in 2009 [27] and it was the foundation for V-NAND (Fig. 11), which is the first 3D memory device who reached the market. Except for SL + regions which are n + diffusions, the equivalent circuit of TCAT is the same of BiCS (Fig. 4). All SL + lines are connected together to form the common Source Line. There are 2 metal layers for decoding wordlines and NAND strings, respectively.

TCAT is based on *gate-replacement* [27], whereas BiCS is *gate-first*. Gate-replacement begins with the deposition of multiple oxide/nitride layers. After the stack formation, nitride is removed through an etching process. Afterwards, tungsten metal gates are deposited and, finally, gates are separated by using another etching step. Metal gates translate into a lower wordline parasitic resistance, resulting in faster programming and reading operations.

The bulk erase operation is another significant difference compared to BiCS. Because NAND strings are close to n + areas, during erasing, holes can come straight

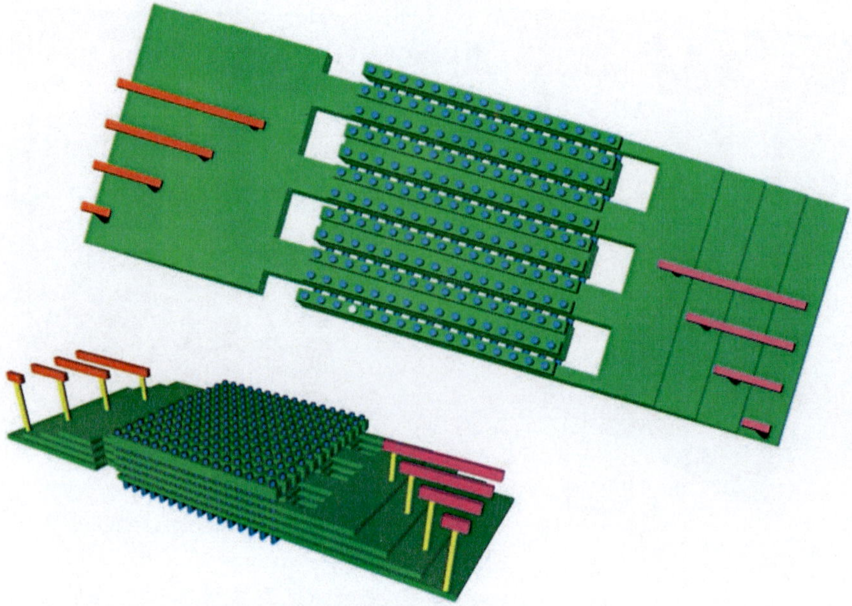

Fig. 9 Fork-shaped fan-out. Adapted with permission from [19]. ©2017 IEEE

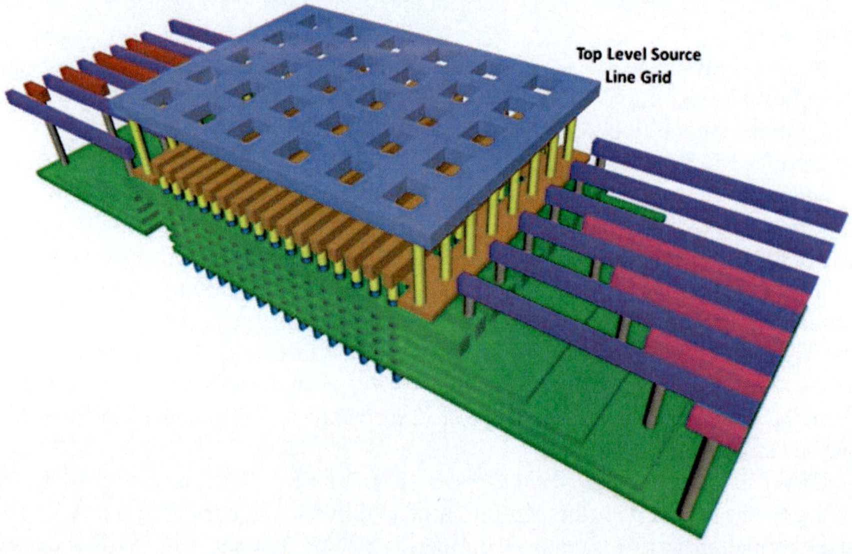

Fig. 10 P-BiCS: Source line metal mesh. Adapted with permission from [19]. ©2017 IEEE

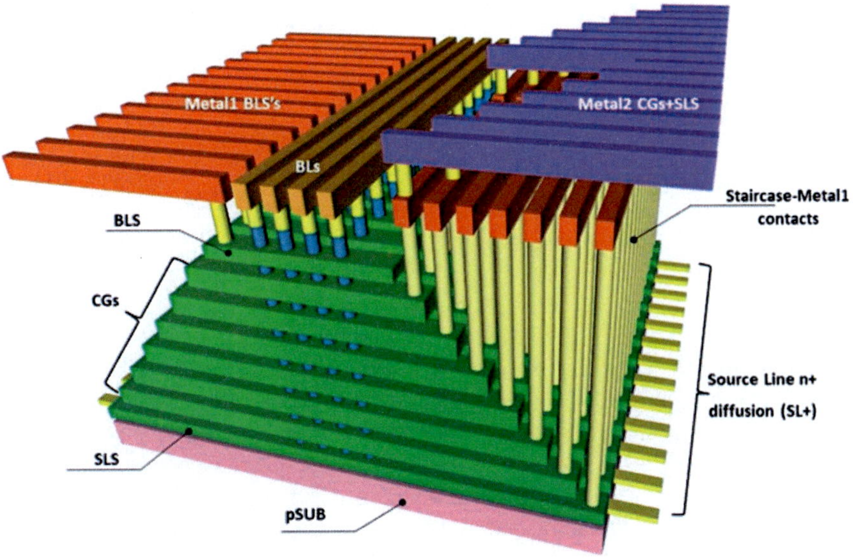

Fig. 11 TCAT NAND flash array

from the substrate, thus avoiding the GIDL (*Gate Induced Drain leakage*) on the source side, which is a well-known problem for BiCS.

BiCS and TCAT are compared in Fig. 12 [28]. Being TCAT based on a gate-last process, the charge trap layer is biconcave, and thanks to this particular shape it is much harder for charges to spread out. On the contrary, BiCS is characterized by a

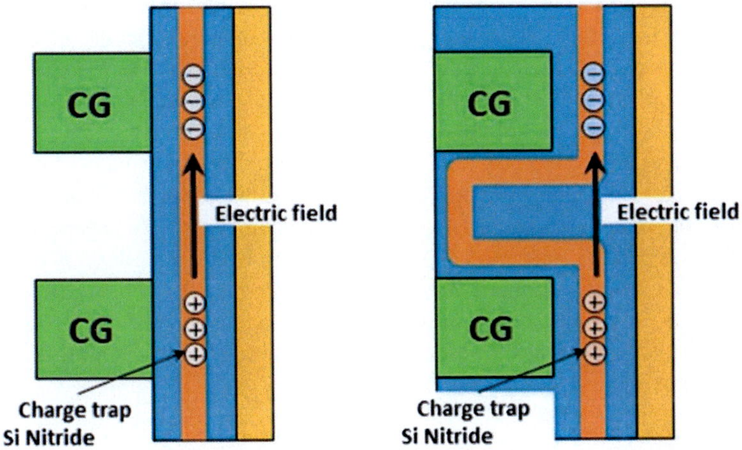

Fig. 12 BiCS versus TCAT

charge trapping layer going through all gate plates, thus acting as a charge spreading path: of course, the main consequence of this layout is a degradation of data retention.

TCAT evolved into another architecture called V-NAND [29]. The first generation had 24 wordline layers, plus additional dummy wordline layers (dummy CG) [30–32].

Why dummy layers? Mainly because of the floating body of the memory cells with vertical channel. In fact, during the programming operations, hot carriers are generated by the high lateral electric field located at the edge of the NAND string. Therefore, these hot carriers keep the voltage on the channel low during the programming operation of the first wordline (i.e. Program Disturb) [33, 34]. Dummy wordlines before the first WL are an effective and simple solution to this problem [35, 36].

A 128 Gb TLC (3 bit/cell) device manufactured by using V-NAND Gen2 was published in 2015 [37, 38]. Gen2 had 32 memory layers instead of the previous 24 and introduced the concept of Single-Sequence Programming. Conventional (mainly 2D) TLC programming techniques go through the programming sequence multiple times. To be more specific, each wordline is programmed 3 times, such that V_{TH} distributions can be progressively tightened. Because of the smaller cell-to-cell interference (compared to FG), CT cells exhibit an intrinsic narrower native V_{TH} distribution. As a result, V-NAND Gen2 could write 3 pages of logic data in a single programming sequence. There are 2 benefits to this approach: reduced power consumption and faster programming.

V-NAND Gen3 appeared in 2016 [39], in the form of a 48 layer TLC device. With such a high number of gate layers, the very high aspect ratio of the pillar becomes a serious challenge for the etching technology. To mitigate this problem, the easiest solution is to shrink the thickness of gate layers. The downside of this approach is that the parasitic RC of the wordline gets higher, thus slowing access operations to the memory array. Moreover, channel's size fluctuations become critical. Indeed, pillars are holes drilled in the gate layer and they represent a barrier for charges flowing along the wordline: in essence, a distribution of the holes diameters generates a distribution of the parasitic resistances of gate layers. In addition, pillars, once manufactured, have the conic shape sketched in Fig. 13. The overall result is that the same voltage applied to different gate layers translates into a waveform per layer. An adaptive program pulse scheme can fix the problem. In a nutshell, the program pulse duration has to be tailored to the characteristics of the wordline layer. As the number of layers increases, the pillar becomes longer with a negative impact on the aspect ratio of the pillar. To compensate for that, V-NAND Gen4 [40], which is built on a stack of 64 layers, had to shrink both the layer thickness and the intra-layer distance (spacing). The downside is an increased wordline parasitic capacitance which adversely affects cell's reliability and timings. Improved circuits and programming algorithms can be used to tackle this problem [40].

As discussed, both BiCS [41] and V-NAND use CT cells, but Floating Gate still exists, as explained in the next Section.

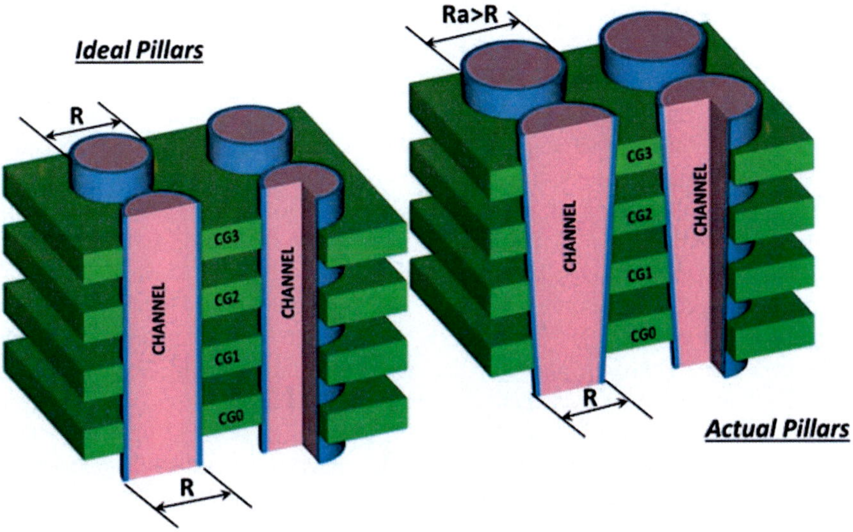

Fig. 13 Ideal versus actual shape of pillars

2 3D Floating Gate NAND Flash Memories

2D NAND Flash memories use FG cells which have been improved and optimized for decades. Of course, there have been many attempts to reuse this know-how in 3D.

The first 3D attempt is known as *3D Conventional FG* (C-FG) or S-SGT (*Stacked-Surrounding Gate Transistor*) [42–44], and it is sketched in Fig. 14.

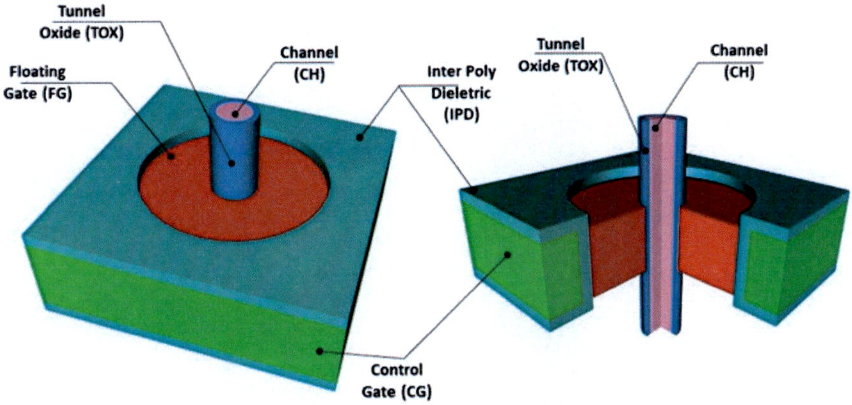

Fig. 14 3D C-FG cell. Adapted with permission from [19]. ©2017 IEEE

A C-FG NAND string is shown in Fig. 15, including select transistors. Please note that both string selectors are manufactured as standard transistors, i.e. they haven't any floating gate. Figure 16 shows a C-FG array, including the fan-out region. While all wordlines at the same height of the stack are connected, BLS lines can't, because they need to be page selective per each CG layer. On the contrary, SLS transistors can be shorted together, thus saving both power and silicon area.

Because we are talking about FG cells, FG coupling between neighboring cells is the main hurdle for vertical scaling. With enhancement-mode operations, the high resistance of source/drain (S/D) regions should also be carefully considered. In fact, these regions need high-doping and this is not very easy to accomplish when the conduction channel is made of polysilicon. The solution to this problem is to electrically invert the S/D layer by using higher voltages during read. This simple solution is hardly manageable by C-FG cells because of the thin FG.

The *Extended Sidewall Control Gate* (ESCG) structure, Fig. 17 [45], is another FG option and it was developed to contain the interference effect. Moreover, by applying a positive voltage to the ESCG structure, density of electrons on the surface of the pillar can be much higher than C-FG (even one order of magnitude): a highly inverted electrical source/drain can significantly lower the S/D resistance.

In addition, the ESCG shielding structure reduces the FG–FG coupling capacitance: the ESCG region is biased as CG, and the CG coupling capacitance (C_{CG}) is significantly increased because of the increased overlap area between CG and FG. A higher CG coupling ratio is one of the key ingredients for achieving effective NAND Flash operations [46].

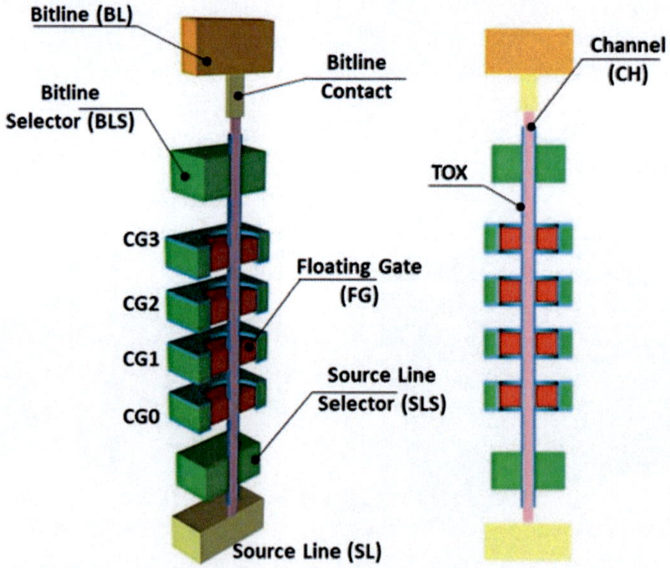

Fig. 15 C-FG NAND flash string. Adapted with permission from [19]. ©2017 IEEE

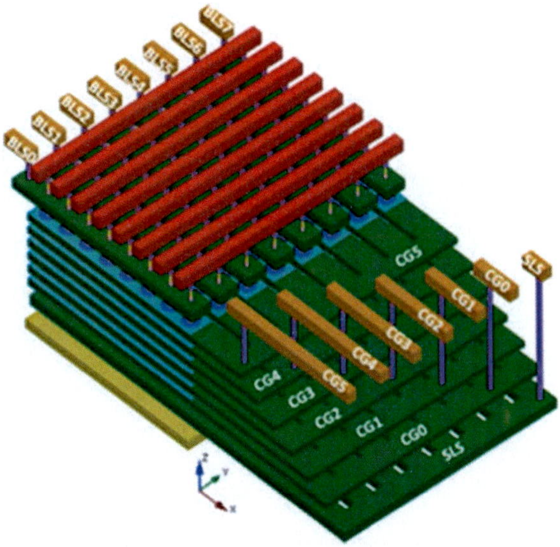

Fig. 16 C-FG NAND flash array with fan-out

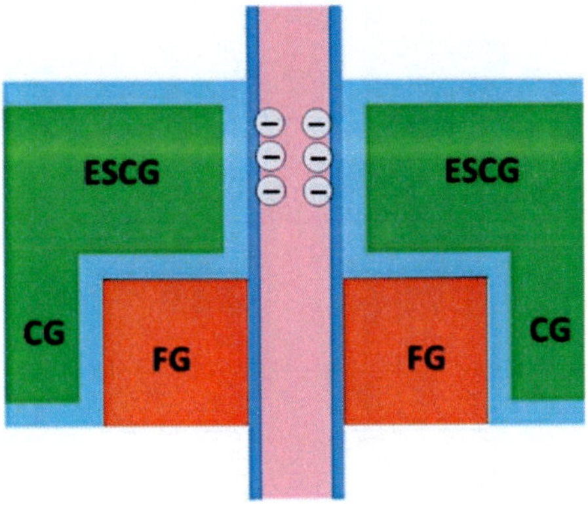

Fig. 17 ESCG NAND flash cell

Another FG cell is DC-SF (*Dual Control-Gate with Surrounding Floating Gate,* Fig. 18) [47]. This time FG is controlled by two CGs. The impact on the FG/CG coupling ratio is remarkable, thanks to the enlargement of the FG/CG overlap area. Another positive aspect is the reduction of the voltages required for programming and erasing. DC-SF eliminates the FG-FG interference because the CG between two adjacent FGs plays the role of an electrostatic shield [48].

FG is fully isolated by IPD (*Inter Poly Dielectric*) and capacitive coupled to upper and lower control gates, CGU and CGL, respectively. The tunnel oxide is located between the channel CH and FG, while IPD is on the sidewall of the CG. In this way, free charges cannot tunnel to the control gates.

BiCS and DC-SF NAND strings are sketched in Fig. 19. In BiCS the nitride layer, going across all gates, makes the cell prone to data retention issues [49]. On

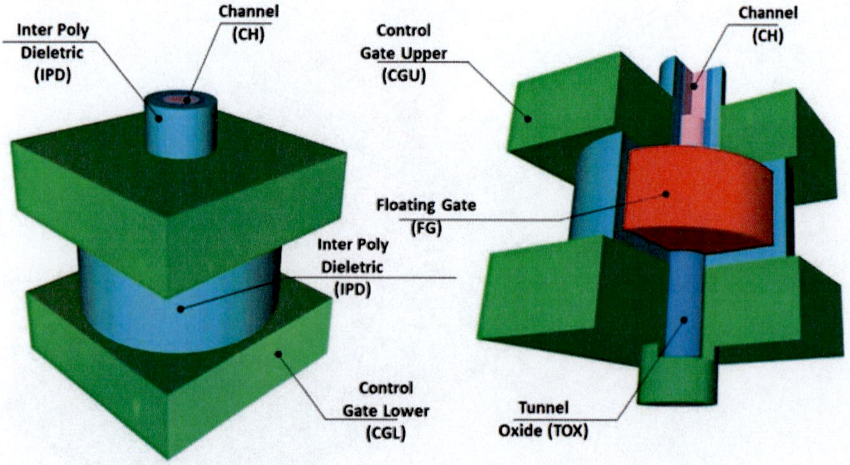

Fig. 18 DC-SF NAND flash cell. Adapted with permission from [19]. ©2017 IEEE

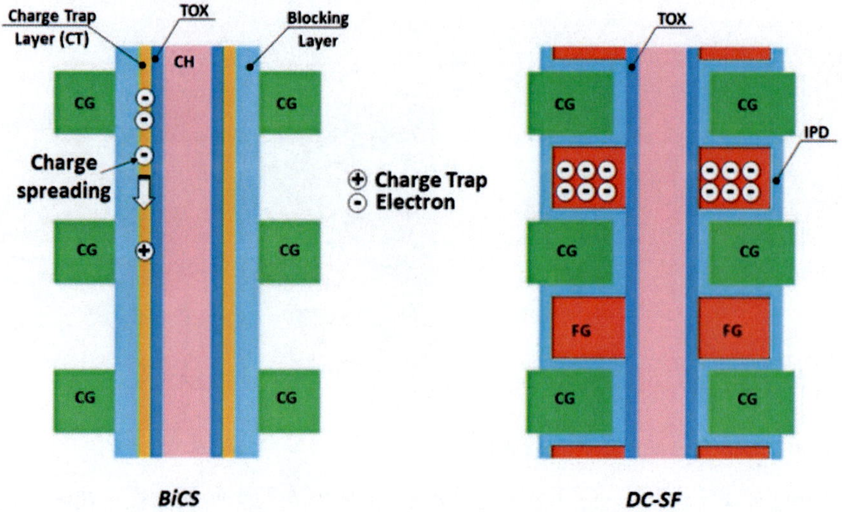

Fig. 19 BiCS versus DC-SF

the contrary, the surrounding FG is totally isolated: it is much easier for DC-SF to retain electrons [50, 51]. Of course, the downside of DC-SF is the fact there are two gate layers instead of one, coupled with much more complex biasing schemes [52, 53].

The *Separated Sidewall Control Gate* (S-SCG) Flash cell [54] displayed in Fig. 20 is another 3D FG option developed around the sidewall concept.

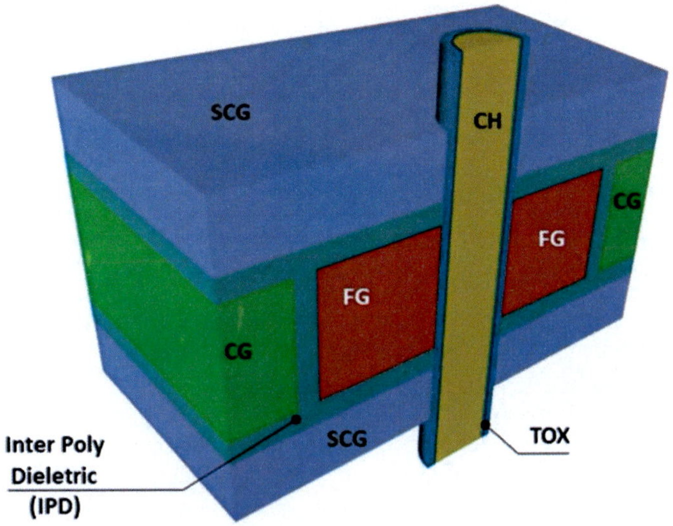

Fig. 20 S-SCG NAND flash cell. Adapted with permission from [19]. ©2017 IEEE

One of major drawbacks of this cell is the "direct" disturb to the neighboring passing cells, caused by the high SCG/FG coupling capacitance. We define it as "direct" because the sidewall CG is shared between adjacent cells: as a matter of fact, biasing SCG means biasing both FGs.

To minimize the decoding complexity, all SCGs belonging to one block adopt a common SCG scheme; besides their electrostatic shield functionality, sidewall gates can help all memory operations [55]. For instance, the common SCG is biased at 1 V during read operations, thus electrically inverting the channel (same as ESCG). Compared to ESCG, the electrical inversion happens simultaneously on source and drain, exactly because of the sidewall gates. Same thing happens during programming: the common SCG is biased at a medium voltage to improve the channel boosting efficiency.

Besides the direct disturb, another problem of Sidewall Gates is the limitation of vertical scaling to around 30 nm; indeed, the thicknesses of SCG and IPD can't be scaled too much, otherwise they would breakdown when voltages are applied.

Let's now take a look at examples of 3D FG NAND memory arrays of hundreds of Gb. The first 3D FG device was published in 2015 [56], in the form of a 384 Gb TLC NAND based on C-FG. This memory device was built with a stack of 32 (+ dummy) memory layers.

A 768 Gb 3D FG NAND became public in the following year [57]. What is unique in this case is the fact that the area underneath the array was used for circuitry. More details about this approach are provided in Sect. 3.

3 Key Challenges for 3D Flash Development

In this Section we cover some of the key challenges that technologists and designers are facing to push 3D memories even further.

3.1 Number of Layers

To reduce the bit size, the number of stacked cells needs to go up, but this causes a bunch of problems hard to solve [6].

Pillar's *Aspect Ratio* (AR) is definitely the first challenge to overcome; in a stack of 32 cells AR can already be as high as 30. In this context, hole etching and gate patterning are extremely difficult, but of paramount importance.

A possible solution to this problem is to divide the stacking process in more steps to reduce the corresponding AR. For example, a NAND string made of 128 cells can be divided in 2 groups of 64 cells each, as shown in Fig. 21. The downside of this solution is the cost of the stacking process (in this example, 4 times higher than the cost of the plain solution).

Second problem is the small cell current [58]. With 2D sensing schemes, a 200 nA/cell saturation current is considered the right value because it gives a reasonable sensing margin. Unfortunately, already with a stack of 24 layers, the cell current is just

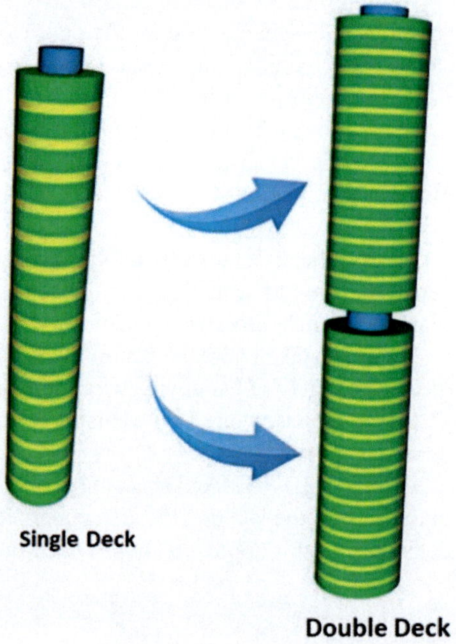

Fig. 21 Multi-stacked or multi-deck process [6]

~20% of FG cell. And it becomes lower and lower as the number of cells in the vertical stack increases. There are a couple of possible paths to solve this problem: sensing schemes with higher sensitivity, and the introduction of new materials enabling a higher cell mobility in the poly-Si channel (i.e. a higher current) [59–62].

All the above-mentioned problems can be fixed if entire NAND strings could be stacked one on top of each other. In this case, either bitlines or source lines are fabricated between NAND strings. This special architecture can simultaneously reduce the aspect ratio and increase the sensing current at same time.

3.2 *Peripheral Circuits Under Memory Arrays*

In the first 3D generations [63, 64], peripheral circuits (charge pumps, logic, etc.) and core circuits (like Page Buffers and Row decoders) are located outside the memory matrix, like in a conventional 2D chip floorplan, as sketched in Fig. 22. However, 3D memory cells are vertically stacked: in other words, memory transistors are not formed on the Si substrate; on the contrary, they are built around a deposited poly-Si (vertical pillar). Therefore, 3D architectures allow placing some circuits directly on the Si substrate under the memory array. Of course, this solution offers a significant reduction of the chip size.

Figure 23 shows a layout of a Flash memory with *Circuits Under the Array* (CuA) [65, 66].

This big area saving doesn't come for free. The most important challenge is manufacturing low resistance metal layers under the array: this is absolutely critical for a reliable circuit functionality. Usually, metal layers used in 2D NAND flash memories are made of Cu. However, when circuits are under the array, the high temperature

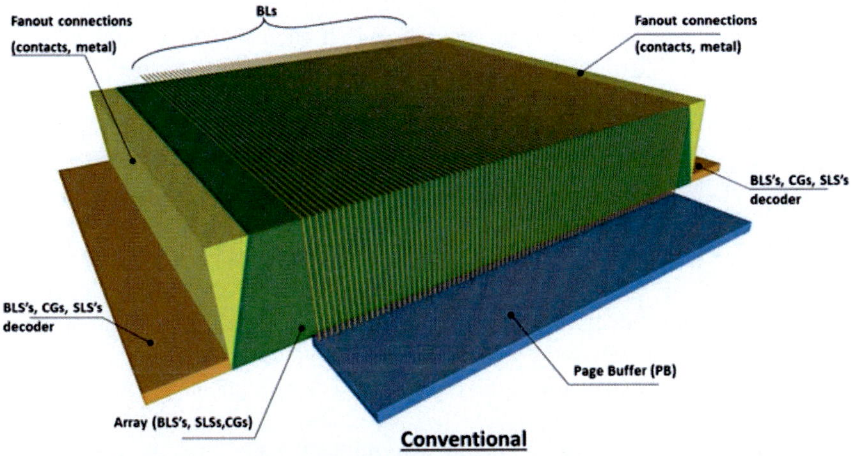

Fig. 22 Conventional 3D NAND flash memory layout

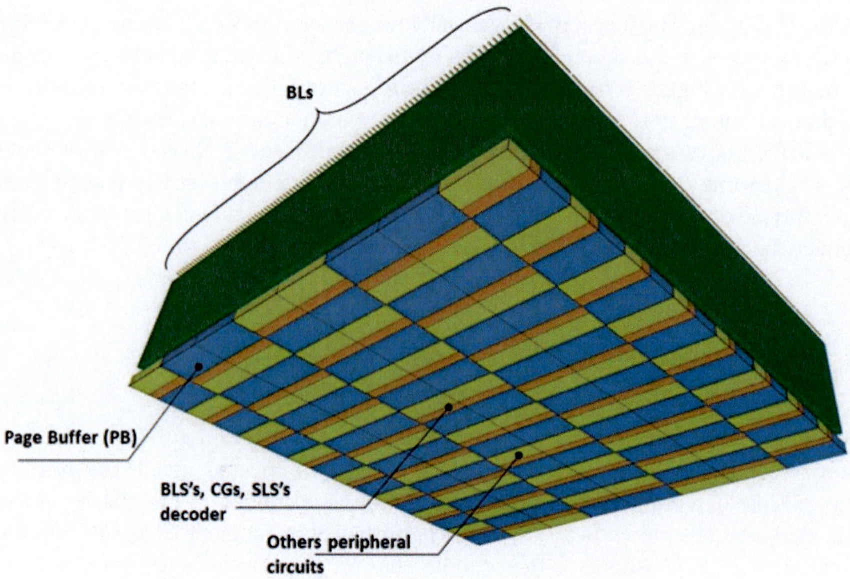

Fig. 23 3D NAND flash memory layout with Circuits Under Array

processes (i.e. >800 °C) that 3D requires can seriously degrade the resistance of metal layers. Therefore, circuits under the array require 3D "low" temperature fabrication processes.

4 Future Trend for 3D NAND Flash

Figure 24 shows cell's size scaling trend, based on published die photographs. 2D became flat below 20 nm, while 3D cell showed a significant reduction going from 24 to 64 layers. This 3D scaling speed will continue by increasing the height of the memory stack, and exploiting technological innovations like Multi-stacked and Stacked NAND string [67].

3D NAND arrays based on CT vertical channel were selected for volume production because the fabrication process is simpler than other 3D architectures. Volume production of 3D NAND Flash started in late 2013 with a 24 layer MLC (2 bit/cell) V-NAND [63, 68]. Year after year, the number of stacked cells grew up, as shown in [7, 64, 69], thus reducing the cost per bit and fueling an even more pronounced diffusion of Solid State Drives.

In this chapter we have presented many architectural options for building a 3D NAND array, including some of the latest and greatest layout options, but the 3D evolution is just at the beginning. In fact, two fundamentally different technologies, Floating and Charge Trap, are fighting each other, trying to prove that they can win

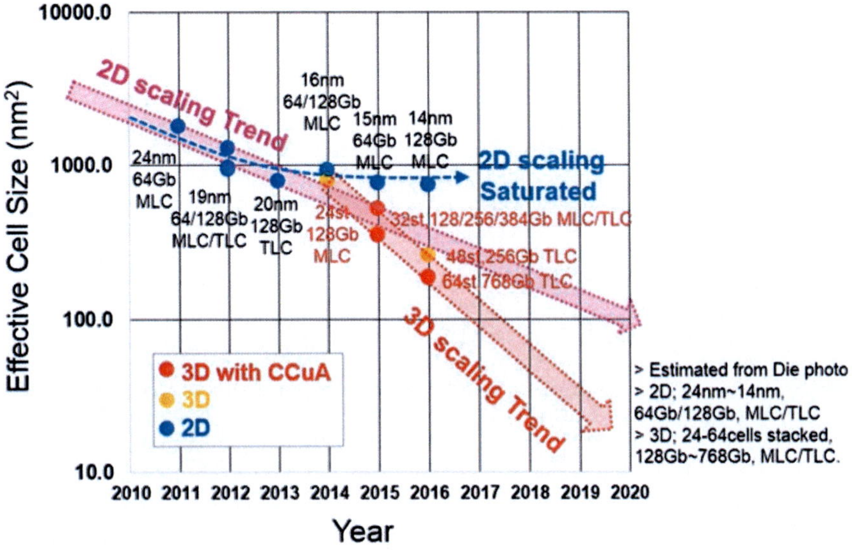

Fig. 24 Effective cell size trend. Reproduced with permission from [19]. ©2017 IEEE

in the long run, i.e. when scaling will be pushed to the limit. Flash manufactures are already shooting for 200 vertical layers with multi-level capabilities, including 4 bit/cell and 5 bit/cell. No doubt that we'll see a lot of innovations in the near future: engineers and scientists are called to give their best effort to make this vertical evolution happen.

References

1. F. Masuoka, M. Momodomi, Y. Iwata, R. Shirota, New ultra high density EPROM and flash EEPROM with NAND structure cell, electron devices meeting. International **33**, 552–555 (1987)
2. R. Micheloni, L. Crippa, A. Marelli, Inside NAND Flash Memories, Chap. 6 (Springer, 2010)
3. T. Mizuno et al., Experimental study of threshold voltage fluctuation due to statistical variation of channel dopant number in MOSFET's. IEEE Trans. Electron Devices **41**(11), 2216–2221 (1994)
4. H. Kurata et al., The impact of random telegraph signals on the scaling of multilevel flash memories, in *Symposium on VLSI Technology* (2006)
5. C.M. Compagnoni et al., Ultimate accuracy for the NAND flash program algorithm due to the electron injection statistics. IEEE Trans. Electron Devices **55**(10), 2695–2702 (2008)
6. S. Aritome, NAND Flash Memory Technologies. IEEE Press Series on Microelectronics System, (Wiley-IEEE Press, 2015)
7. S. Aritome, 3D Flash Memories, International Memory Workshop 2011 (IMW 2011), short course
8. R. Micheloni, L. Crippa, A. Marelli, Inside NAND Flash Memories, Chap. 5 (Springer, 2010)
9. http://www.samsung.com/us/business/oem-solutions/pdfs/VNAND_technology_WP.pdf. Samsung V-NAND technology, White Paper, 2014

10. R. Micheloni, L. Crippa, Chapter 3, Multi-bit NAND flash memories for ultra high density storage devices, in *Advances in Non-volatile Memory and Storage Technology*, ed. by Y. Nishi (Woodhead Publishing, Sawston, 2014)
11. R. Micheloni et al., Chapter 7, High-capacity NAND flash memories: XLC storage and single-die 3D, in *Memory Mass Storage*, ed. by G. Campardo et al. (Springer, 2011)
12. H. Tanaka et al., Bit cost scalable technology with punch and plug process for ultra high density flash memory, in *VLSI Symposium Technical Digest* (2007), pp. 14–15
13. Y. Fukuzumi et al., Optimal integration and characteristics of vertical array devices for ultra-high density, bit-cost scalable flash memory, in *IEDM Technical Digest* (2007), pp. 449–452
14. M. Ishiduki et al., Optimal device structure for pipe-shaped BiCS flash memory for ultra high density storage device with excellent performance and reliability, in *IEDM Technical Digest* (2009), pp. 625–628
15. T. Maeda et al., Multi-stacked 1G cell/layer pipe-shaped BiCS flash memory, in *Digest Symposium on VLSI Circuits* (2009), pp. 22–23
16. R. Katsumata et al., Pipe-shaped BiCS flash memory with 16 stacked layers and multi-level-cell operation for ultra high density storage devices, in *Symposium on VLSI Technology* (2009), pp. 136–137
17. H.-T. Lue, S.-Y. Wang, E.-K. Lai, K.-Y. Hsieh, R. Liu, C.Y. LuA, BESONOS (Bandgap Engineered SONOS) NAND for Post-Floating Gate Era Flash Memory, in *Symposium on VLSI Technology* (2007)
18. H. Aochi, BiCS flash as a future 3-D non-volatile memory technology for ultra high density storage devices, in *Proceedings of International Memory Workshop* (2009), pp. 1–2
19. R. Micheloni, S. Aritome, L. Crippa, Array architectures for 3-D NAND flash memories. Proc. IEEE **105**(9), 1634–1649 (2017). https://doi.org/10.1109/JPROC.2017.2697000
20. Y. Yanagihara et al., Control gate length, spacing and stacked layers number design for 3D-Stackable NAND flash memory2, in *IEEE IMW* (2012), pp. 84–87
21. K. Takeuchi, Scaling challenges of NAND flash memory and hybrid memory system with storage class memory and NAND flash memory, in *IEEE Custom Integrated Circuits Conference (CICC)* (2013), pp. 1–6
22. A. Nitayama et al., Bit Cost Scalable (BiĊS) flash technology for future ultra high density storage devices, in *International Symposium on VLSI Technology Systems and Applications (VLSI TSA)*, (2010), pp. 130–131
23. Y. Komori et al., Disturbless flash memory due to high boost efficiency on BiCS structure and optimal memory film stack for ultra high density storage device, in *IEDM Technical Digest* (2008), pp. 851–854
24. J. Kim et al., Novel 3-D structure for ultra high density flash memory with VRAT (vertical-recess-array-transistor) and PIPE (planarized integration on the same plane), in *IEEE Symposium on VLSI Technology* (2008)
25. J. Kim et al., Novel vertical-stacked-array-transistor (VSAT) for ultra-high-density and cost-effective NAND flash memory devices and SSD (solid state drive), in *IEEE Symposium on VLSI Technology* (2009)
26. H.T. Lue, T.H. Hsu et al., A highly scalable 8-layer 3D vertical-gate (VG) TFT NAND flash using junction-free buried channel BE-SONOS device, in *VLSI Symposia on Technology* (2010)
27. J. Jang et al., Vertical cell array using TCAT (terabit cell array transistor) technology for ultra high density NAND flash memory, in *IEEE Symposium on VLSI Technology* (2009)
28. W. Cho et al., Highly reliable vertical NAND technology with biconcave shaped storage layer and leakage controllable offset structure, in *Symposium on VLSI Technology (VLSIT)* (2010), pp. 173–174
29. J. Elliott, E.S. Jung, Ushering in the 3D memory era with V-NAND, in *Proceedings of Flash Memory Summit*, (Santa Clara, CA, 2013), www.flashmemorysummit.com
30. K.-T. Park, Three-dimensional 128 Gb MLC vertical NAND flash memory with 24-WL stacked layers and 50 MB/s high-speed programming, in *IEEE ISSCC, Digest Technical Papers* (2014), pp. 334–335

31. K.-T. Park, Three-dimensional 128 Gb MLC vertical NAND flash memory with 24-WL stacked layers and 50 MB/s high-speed programming. IEEE J. Solid-State Circ. **50**(1), (2015)
32. K.T. Park, A world's first product of three-dimensional vertical NAND flash memory and beyond, in *NVMTS* (27–29 October 2014)
33. K.-S. Shim et al., Inherent issues and challenges of program disturbance of 3D NAND flash cell, in *Memory Workshop (IMW), 2012 4th IEEE International* (20–23 May 2012), pp. 1, 4
34. H.S. Yoo et al., Modeling and optimization of the chip level program disturbance of 3D NAND Flash memory, in *Memory Workshop (IMW), 2013 5th IEEE International* (26–29 May 2013), pp. 147, 150
35. E. Choi et al., Device considerations for high density and highly reliable 3D NAND flash cell in near future, in *IEEE International Electron Devices Meeting* (2012), pp. 211–214
36. K. Shim et al., Inherent issues and challenges of program disturbance of 3D NAND flash cell, in *IEEE International Memory Workshop* (2012), pp. 95–98
37. J.-W. Im, 128 Gb 3b/cell V-NAND flash memory with 1 Gb/s I/O rate, in *IEEE International Solid-State Circuits Conference* (2015), pp. 130–131
38. J.-W. Im, 128 Gb 3b/cell V-NAND flash memory with 1 Gb/s I/O rate. J. Solid-State Circuits. **51**(1), (2016)
39. D. Kang et al., 256 Gb 3b/Cell V-NAND flash memory with 48 stacked WL layers, in *IEEE International Solid-State Circuits Conference (ISSCC), Digest Technical Papers* (2016), pp. 130–131
40. C. Kim et al. A 512 Gb 3b/cell 64-stacked WL 3D V-NAND flash memory, in *Solid-State Circuits Conference Digest of Technical Papers (ISSCC), 2017 IEEE International* (2017) pp. 202–203
41. R. Yamashita et al., A 512 Gb 3b/cell Flash Memory on 64-Word-Line-Layer BiCS Technology, in *Solid-State Circuits Conference Digest of Technical Papers (ISSCC), 2017 IEEE International* (2017), pp. 196–197
42. T. Endoh et al., Novel ultra high density flash memory with a stacked-surrounding gate transistor (S-SGT) structured cell. IEDM Tech. Dig. 33–36 (2001)
43. T. Endoh et al., Novel ultra high density flash memory with a stacked-surrounding gate transistor (S-SGT) structured cell.2. IEEE Trans. Electron Devices **50**(4), 945–951 (2003)
44. T. Endoh et al., Floating channel type SGT flash memory, in *The 1999 Joint International Meeting*, Hawaii, vol. 99–2, Abstract No. 1323, 17–22 Oct 1999
45. M.S. Seo et al., The 3-dimensional vertical FG nand flash memory cell arrays with the novel electrical S/D technique using the extended sidewall control gate (ESCG), in *Proceedings of IEEE International Memory Workshop* (2010), pp. 1–4
46. M.S. Seo et al., 3-D Vertical FG NAND flash memory with a novel electrical S/D technique using the extended sidewall control gate. IEEE Trans. Electron Devices **58**(9), (2011)
47. S. Whang et al., Novel 3-dimensional dual control gate with surrounding floating-gate (DC-SF) NAND flash cell for 1 Tb file storage application, in *Proceedings of International Electron Devices Meeting (IEDM)* (2010), pp. 668–671
48. Y. Noh et al., A new metal control gate last process (MCGL process) for high performance DC-SF (dual control gate with surrounding floating gate), in *3D NAND flash memory in Symposium on VLSI Technology* (2012), pp. 19–20
49. C.-P. Chen et al., Study of fast initial charge loss and its impact on the programmed states Vt distribution of charge-trapping NAND Flash, in *Electron Devices Meeting (IEDM), 2010 IEEE International* (6–8 Dec 2010), pp. 5.6.1, 5.6.4
50. R. Micheloni, L. Crippa, Multi-bit NAND flash memories for ultra high density storage devices (chapter 3), in *Advances in Non-volatile Memory and Storage Technology*, ed, by Y. Nishi (Woodhead Publishing, 2014)
51. R. Micheloni et al., High-capacity NAND flash memories: XLC storage and single-die 3D (chapter 7) in *Memory Mass Storage*, eds. by G. Campardo et al. (Springer, 2011)
52. H. Yoo et al., New read scheme of variable Vpass-read for dual control gate with surrounding floating gate (DC-SF) NAND flash cell, in *Proceedings of 3rd IEEE International Memory Workshop* (2011), pp. 1–4

53. S. Aritome et al., Advanced DC-SF cell technology for 3-D NAND flash. IEEE Trans. Electron Devices **60**(4), 1327–1333 (2013)
54. M.S. Seo et al., A novel 3-D vertical FG nand flash memory cell arrays using the separated sidewall control gate (S-SCG) for highly reliable MLC operation, in *Proceedings of 3rd IEEE International Memory Workshop (IMW)* (2011), pp. 1–4
55. M.S. Seo et al., Novel concept of the three-dimensional vertical FG nand flash memory using the separated-sidewall control gate. IEEE Trans. Electron Devices **59**(8), 2078–2084 (2012)
56. K. Parat, C. Dennison, A floating gate based 3D NAND technology with CMOS under array, in *Conference on International Electron Devices Meeting (IEDM)* (San Francisco, USA, 2015)
57. T. Tanaka et al., A 768 Gb 3 b/cell 3D-floating-gate NAND flash memory, in *2016 IEEE International Solid-State Circuits Conference (ISSCC), Digest of Technical Papers* (San Francisco, USA, 2016), pp. 142–143
58. E.-S. Choi, S.-K. Park, Device considerations for high density and highly reliable 3D NAND flash cell in near future, in *Electron Devices Meeting (IEDM), 2012 IEEE International*, vol. no., pp. 9.4.1–9.4.4, 10–13 Dec 2012
59. Subirats et al., Impact of discrete trapping in high pressure deuterium annealed and doped poly-Si channel 3D NAND macaroni, in *IEEE International Reliability Physics Symposium (IRPS)* (2017)
60. L. Breuil, Improvement of poly-Si channel vertical charge trapping NAND devices characteristics by high pressure D2/H2 annealing, in *IEEE 8th International Memory Workshop (IMW)* (2016)
61. E. Capogreco et al. MOVPE In1-xGaxAs high mobility channel for 3-D NAND memory, in *IEEE International Electron Devices Meeting (IEDM)* (2015)
62. J. G. Lisoni et al., Laser Thermal Anneal of polysilicon channel to boost 3D memory performance, in *Symposium on VLSI Technology (VLSI-Technology)*, Digest of Technical Papers (2014)
63. K.-T. Park et al., Three-Dimensional 128 Gb MLC Vertical nand Flash Memory With 24-WL Stacked Layers and 50 MB/s High-Speed Programming. IEEE J. Solid-State Circuits **50**(1), 204–213 (2015)
64. J.-W. Im et al., A 128Gb 3b/cell V-NAND Flash Memory with 1Gb/s I/O rate, in *Solid-State Circuits Conference Digest of Technical Papers (ISSCC), 2015 IEEE International* (2015), pp. 23–25
65. S. Aritome, NAND flash memory revolution, in *2016 IEEE 8th International Memory Workshop (IMW)* (Paris, 2016), pp. 1–4
66. T. Tanaka et al., 7.7 A 768Gb 3b/cell 3D-floating-gate NAND flash memory, in *2016 IEEE International Solid-State Circuits Conference (ISSCC)* (San Francisco, CA, 2016), pp. 142–144
67. R. Micheloni (ed.), *3D Flash Memories* (Springer, 2016)
68. K.-T. Park et al., 19.5 Three-dimensional 128Gb MLC vertical NAND Flash-memory with 24-WL stacked layers and 50MB/s high-speed programming, in *Solid-State Circuits Conference Digest of Technical Papers (ISSCC), 2014 IEEE International* (9–13 Feb 2014) pp. 334–335
69. S. Aritome, Joint Rump session in VLSI Symposium 2012, Scaling challenges beyond 1Xnm DRAM and NAND Flash

Deep Neural Network Engines Based on Flash Technology

Rino Micheloni, Luca Crippa, and Cristian Zambelli

Abstract Artificial Neural Networks based on non-volatile memories can easily beat the equivalent full-CMOS implementation (through logic gates) in terms of both speed and energy consumption. After 30 years of development (Mead in Analog VLSI and neural systems. Addison-Wesley, Reading, MA, USA, 1989), few recent works (Indiveri in Front Neurosci 5:1–23, 2011; Likharev in Sci Adv Mater 3:322–331, 2011; Hasler and Marr in Front Neurosci 7, 2013; Ceze et al. in Proceedings of DRC, Newark, DE, USA, pp. 1–2, 2016) has proven that the analog implementation, by adopting nanoscale devices, can even approach the power efficiency of the human brain. The enabling device of the analog approach is a non-volatile memory cell; in this chapter we show how Flash memories (of both NOR and NAND type) can be used to implement the Vector-by-Matrix (VbM) multiplication, which is the core of the hardware implementation of a Neural Network. Indeed, Flash memory cells, thanks to their tunable threshold voltage, can replicate the behavior of a synapse inside the human brain.

1 Introduction

Artificial Neural Networks based on non-volatile memories can easily beat the equivalent full-CMOS implementation (through logic gates) in terms of both speed and energy consumption. After 30 years of development [1], few recent works [2–5] has proven that the analog implementation, by adopting nanoscale devices, can even approach the power efficiency of the human brain. How is that even possible? From

R. Micheloni (✉) · C. Zambelli
Dipartimento di Ingegneria, Università degli Studi di Ferrara, Via G. Saragat 1, 44122 Ferrara, Italy
e-mail: rino.micheloni@unife.it

C. Zambelli
e-mail: cristian.zambelli@unife.it

L. Crippa
Via Manzoni 66, 20874 Busnago, Italy
e-mail: luca.crippa@ieee.org

R. Micheloni and C. Zambelli (eds.), *Machine Learning and Non-volatile Memories*,
https://doi.org/10.1007/978-3-031-03841-9_6

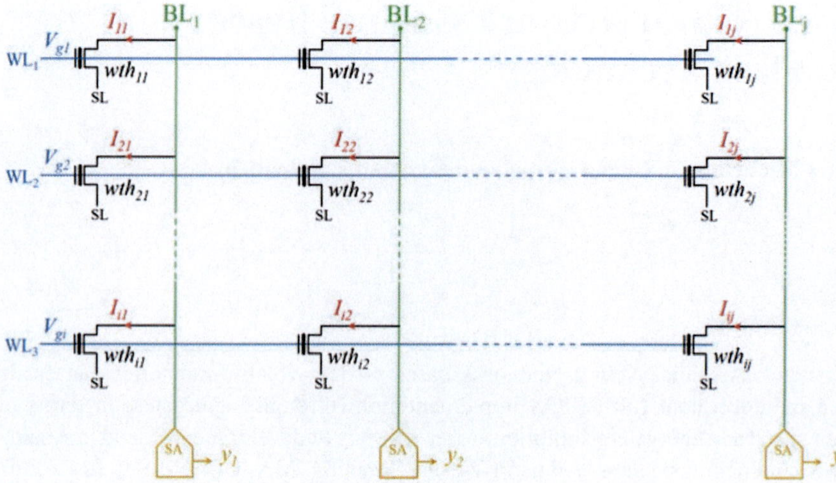

Fig. 1 Analog vector-by-matrix (VbM) multiplication based on floating-gate memory cells

the power consumption point of view, the key operation of the neuromorphic architecture is the Vector-by-Matrix (VbM) multiplication; while in the digital domain it can only be implemented by means of logic gates, in the analog domain it can be compacted in a single structure as shown in Fig. 1. The enabling device of the latter implementation is a non-volatile memory cell (a floating gate transistor in the drawing) which, thanks to its tunable threshold voltage, can replicate the behaviour of a synapse inside the human brain. A bird's-eye view of a floating gate type transistor is reported in Fig. 2, together with its schematic symbol.

Essentially, the VbM implementation based on floating-gate cells can be seen as the evolution of the memristive arrays sketched in Fig. 3.

Researchers sought a replacement of floating gate cells with a plethora of alternative nonvolatile memory devices, from phase change to ferroelectric, from magnetic memories to memristors [6–10]. It is worth highlighting that, so far, their on-chip implementations targeted only small sized deep neural networks or simple neuromorphic tasks [11–15], which means that those technologies need further development, as today's neural networks require a large set of VbM multiplications.

2 Deep Neural Networks Based on NOR Flash Memories

Let's now take a closer look at an implementation of the VbM multiplication based on embedded Flash memories of floating-gate type arranged in a NOR architecture [15–17].

The dependency of the memory cell drain current with respect to the applied gate voltage is depicted in Fig. 4 [1, 4]. In the subthreshold conduction region, we can

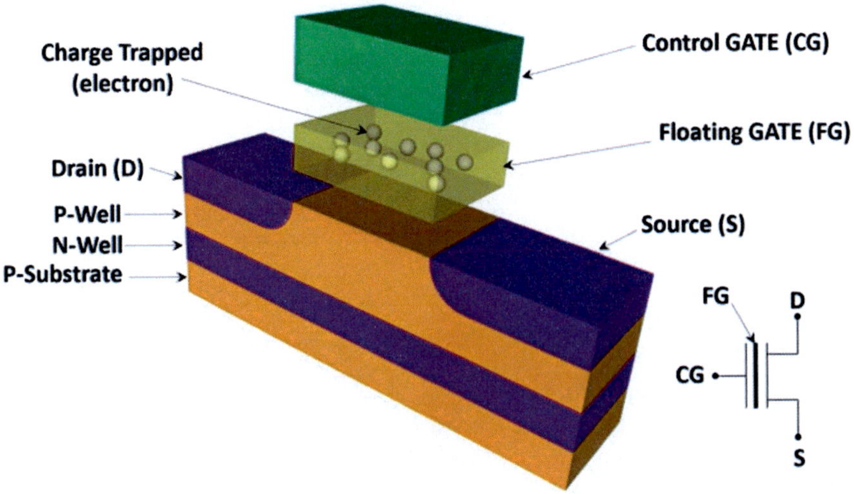

Fig. 2 Floating gate memory cell and its schematic symbol

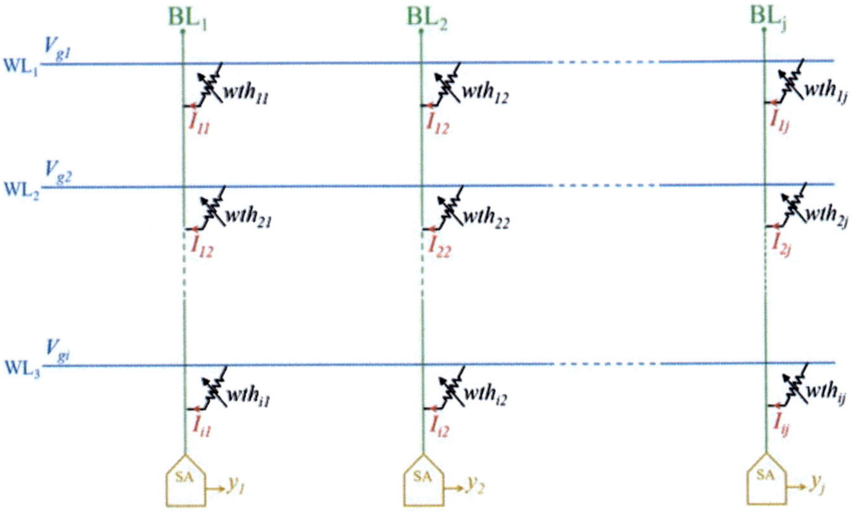

Fig. 3 Analog VbM multiplication circuit with memristive elements

write

$$I_{ij} = I_0 \exp\left(\beta \frac{Vgs_{ij} - Vth_{ij}}{V_T}\right) \tag{1}$$

where

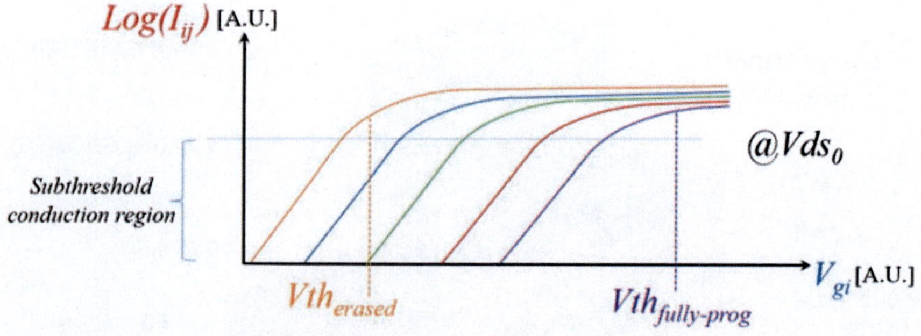

Fig. 4 Drain current as a function of the gate voltage, for several *Vth* memory states

- I_{ij} is the drain current of the *ij* memory cell;
- Vth_{ij} is the threshold voltage of the memory cell (which is a function of its state);
- $VT \sim k_B T/e \sim 26$ mV at room temperature;
- β is the subthreshold slope, which is a function of the gate-to-channel coupling and is usually lower than one.

As described in Chap. 2, at the core of a neural network we have the synaptic weights, whose values are calculated during the training process of the network itself. The most common approach for computing these weights is based on a backpropagation algorithm driven by the target cost function; in essence, goal is to minimize the difference between the correct value and the second largest output of the trained neural network (please refer to Chap. 2). Once the weights are available, their values are used to program the threshold voltage of the memory cells depicted in Fig. 3.

The hardware implementation here described refers to a neural network used as a classifier to be tested with the very well-known MNIST dataset [26]. Binary input is the pixel intensity (i.e., black is logical '0', and white is logical '1').

The architecture of the implemented neural network is sketched in Fig. 5: it is a 3-layer perceptron with a single hidden layer featuring *i* neurons in the input layer, *j* neurons in the hidden layer, and *k* neurons in the output layer. The activation function is the pretty common "sigmoid" function (please refer to Chap. 2 of this book for its mathematical description).

Weights wth_{ij} e wth_{jk} are used to fully-connect adjacent layers according to the standard perceptron architecture. Weights are dialed-in by tuning the threshold voltage of the floating gate memory cells drawn in Fig. 1.

Figure 6 shows the connection between a hidden neuron and the first layer of the network in more details; Fig. 7 depicts the circuit implementation of the same hidden neuron.

The mixed-signal VbM multiplication inside the crossbar memory array makes use of the following voltages:

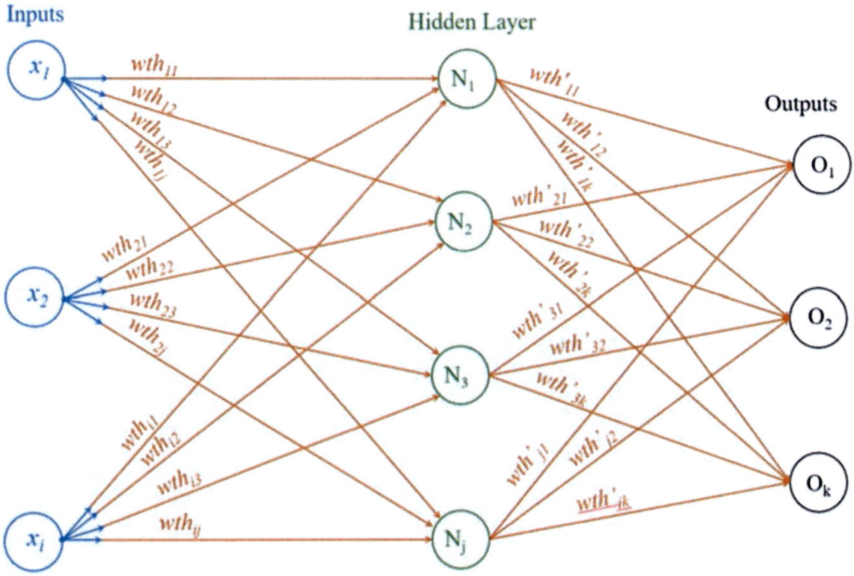

Fig. 5 NOR-based neural network architecture used for hardware implementation

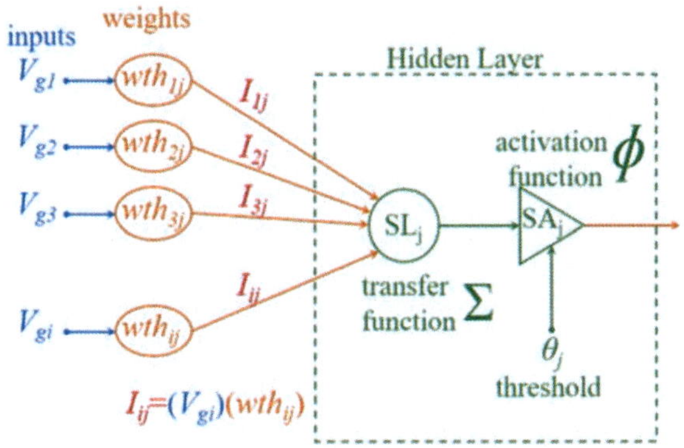

Fig. 6 A generic neuron in the hidden layer *Nj* with its connections to the preceeding input layer

- Input voltage $Vg_i = 4.2$ V for black pixels
- Input voltage $Vg_i = 0$ V for white pixels
- Source line voltage $SL_j = 1.65$ V
- Bitlines voltage $BL_i = 2.7$ V.

Since bitline and source line voltages are kept fixed, cell's current is dictated by its threshold voltage (i.e., weight) and Vg_i only. In other words, the current of the memory cell sitting at the crosspoint of row j and column i does not depend on the conduction state of the surrounding memory cells.

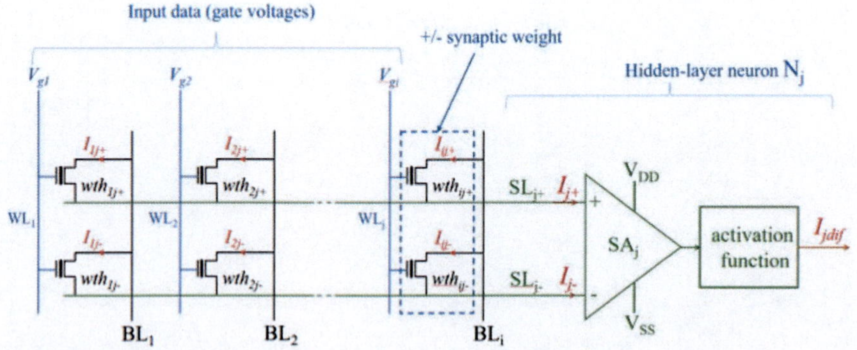

Fig. 7 Circuit implementation of the hidden neuron *Nj* of Fig. 6. The concept of the figure is based on the description provided in [15, 16]

The source line SL is used to connect the source terminal of all the memory cells sitting on the same row and it is biased at a fixed voltage. Thanks to this architecture, the output current *Ij* is exactly the sum of $Vgs_i * wth_{ij}$ over all columns *i*, thus achieving the goal of implementing the VbM multiplication according to the following equation ("bias" is not shown in the circuit):

$$I_j = \sum_{n=1}^{i} Vg_n * wth_{nj} + \text{bias} \tag{2}$$

Another specific aspect of the implementation described in Fig. 7 is the usage of a differential scheme. In fact, each weight is stored in two cells per column instead of one. If the weight is positive, then it is stored in wth_{1j+}, otherwise in wth_{1j-}. Please note that this is a trick to work around the fact that the threshold voltage of a memory cell in the programmed state can only be positive (while the output of the training process of a neural network can be positive or negative). The overall goal here is twofold: to compensate for random drifts on one side and to use zero-centered signals on the other side. Let's now get into the details of how the differential approach works. As described above, there are two memory cells for each synaptic weight: depending on the sign (positive or negative), one of the two cell is programmed to the absolute value of the weight itself, while the other cell's threshold voltage is programmed to the highest possible value, thus being, as a matter of fact, OFF in all working conditions. In this way, only one cell per couple can be ON. Indeed, this is a way to reduce the power consumption (half of the cells are in idle mode) while simplifying the design. Of course, the cost associated with this architecture is the doubling of the overall number of memory cells. With reference to Fig. 7, output currents are then subtracted by using an operational amplifier and the result goes through the circuit that implements the activation function (sigmoid in this case).

Figure 8 describes the connection among the neuron in the output layer O_k and its connections to the hidden layer; Fig. 9 is the corresponding circuit implementation

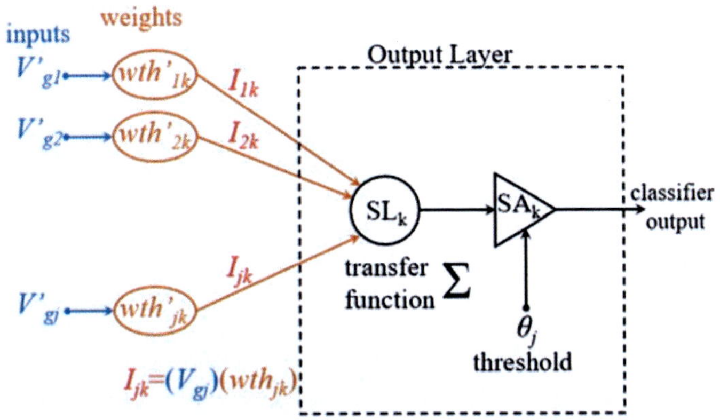

Fig. 8 Generic neuron O_k in the output layer and its connections to the preceding hidden layer

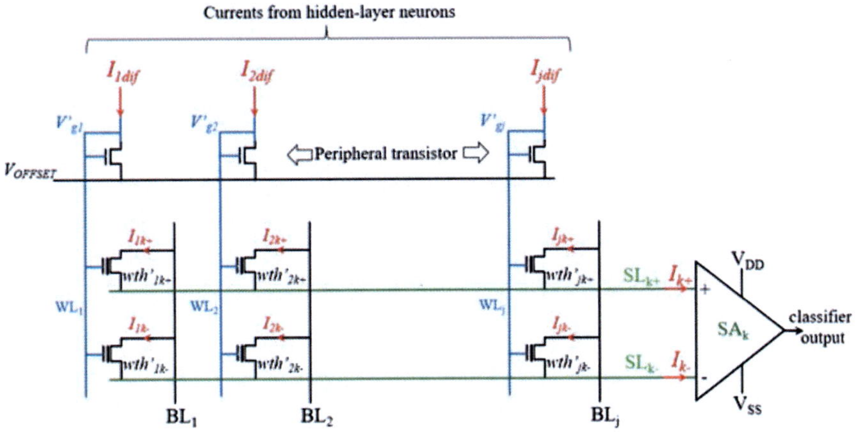

Fig. 9 Circuit implementation of the neuron O_k depicted in Fig. 8. The concept of the figure is based on the description provided in [15, 16]

[18]. At a first sight, the circuit of Fig. 9 is very similar to the circuit of Fig. 7 but there is one key difference: the diode-connected peripheral transistors, which are used to convert the currents coming from the hidden layers into the voltages to be applied to the wordlines. It is worth highlighting that, for matching reasons, the peripheral transistors are identical to memory cells (except for the floating gate).

Output currents *Ijk* can then be summed as in the circuit described in Fig. 7 (with the same differential scheme); therefore, by feeding the circuit of Fig. 9 with the currents I_{jdif} output by the activation function, the full VbM can be realized as described by the following equation ("bias" not shown in the circuit):

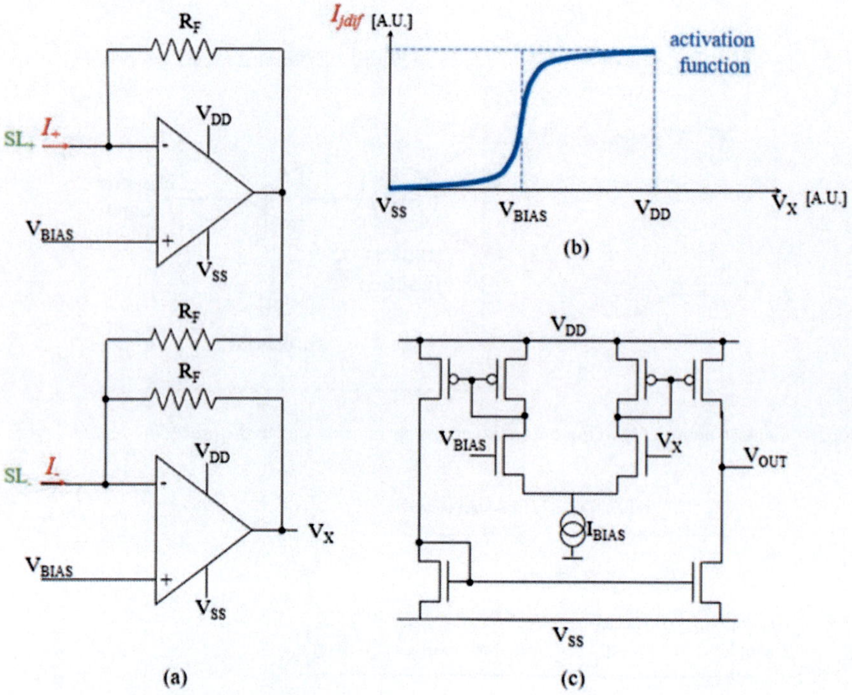

Fig. 10 **a** Differential current amplifier **b** sigmoid function **c** transistor-level schematics of (**b**)

$$I_k = \sum_{n=1}^{j} Vg_j * wth'_{nk} + \text{bias} \tag{3}$$

Figure 10a sketches the circuit used to subtract I_+ and I_- by using two operational amplifiers (in the following simply "op-amp"). Outside the op-amp saturation region, the output voltage V_X can simply be computed as follows:

$$V_X = R_F(I_+ - I_-) + V_{\text{BIAS}} \tag{4}$$

As already mentioned, the activation function selected for the hardware implementation described in this section is the sigmoid function, whose I/V curve is drawn in Fig. 10b; its description at transistor level can be found in Fig. 10c.

The NOR-based VbM multiplication can also be performed by implementing a bitline sensing [19, 20], as sketched in Fig. 11. This time the input signal is the source line (*SL*) voltage (it was the wordline voltage in Fig. 9); bitlines are kept at a fixed voltage, while sense amplifiers are tied to bitlines. Again, sense amplifiers are differential to handle both positive and negative weights. As such, a single synaptic cell is made of two memory cells with 2 drain nodes (BL_{j+}, BL_{j-}) and a single source line (SL_i) in common.

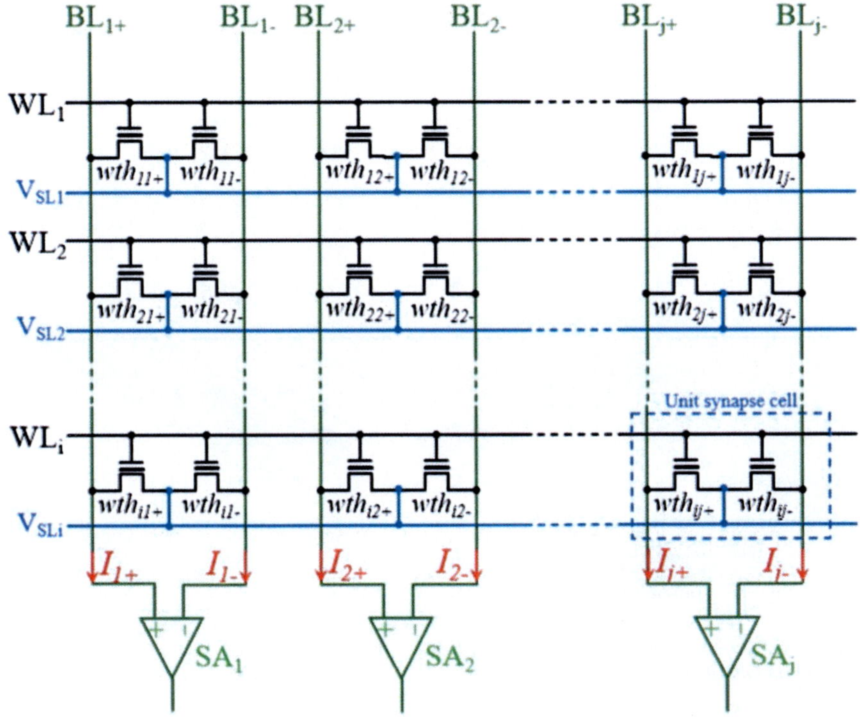

Fig. 11 NOR architecture with bitline sensing. The concept of the figure is based on the description provided in [19, 20]

Currents I_{j+} e I_{j-} can be written as follows

$$I_{j+} = \sum_{n=1}^{i} V_{SLn} * wth_{nj+} \tag{5}$$

$$I_{j-} = \sum_{n=1}^{i} V_{SLn} * wth_{nj-} \tag{6}$$

The output of the sense amplifier is then proportional to the difference $(I_{j+} - I_{j-})$.

Conceptually, this NOR architecture can also be modified to scale in the third dimension, as drawn in Fig. 12, where four 3D layers are shown. Because each layer needs to be independently addressed, an additional (compared to a conventional NOR architecture) decoding stage is required: the Layer Decoder. The role of the layer decoder (essentially, a bunch of transistors used as pass-transistors) is to transfer the voltage coming from the global wordline decoder to the wordlines of the selected layer only. To avoid floating wordlines, the GND Decoder takes care of all the wordlines sitting on unselected 3D layers.

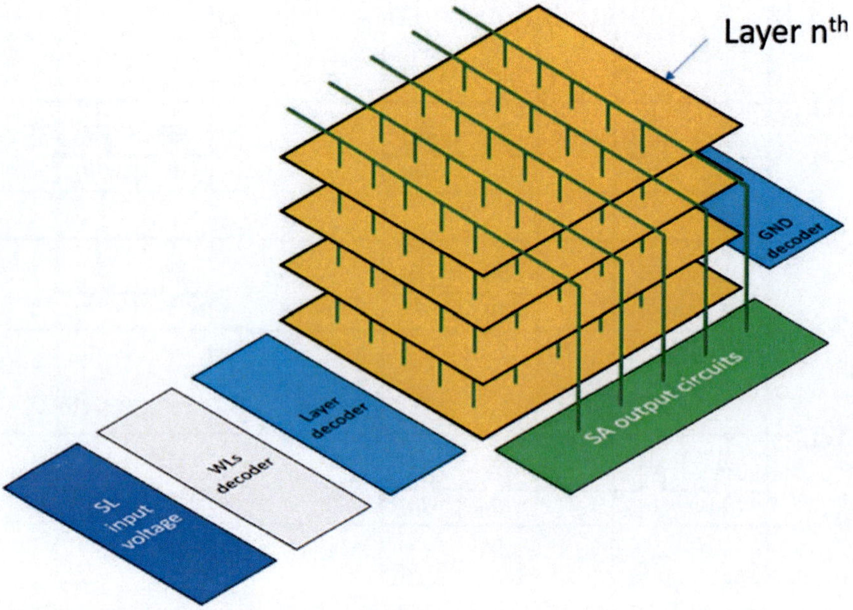

Fig. 12 Implementation of a multi-layer structure using the 3D NOR concept based on the description provided in [19, 20]

While for NOR Flash memories 3D integration is still on paper, most of the NAND memories sold in the market today are 3D. Therefore, it is key to understand how 3D NAND can be used to implement the VbM multiplication.

3 Deep Neural Networks Based on NAND Flash Memories

In Chap. 2 we have seen what Neural Networks (NNs) are, how they work, and a few examples of their practical applications. In essence, they mimic the human brain by adopting a neuron/synapse architecture, which translates into being able of performing parallel data processing. The first neuron implementation was the Perceptron model in 1958 which is sketched in Fig. 13. This section describes a hardware implementation of a perceptron-based neural network by using a NAND Flash memory [21], of a Floating Gate (FG) type.

Each input x_n represents a synapse and w_{nj} is called the "synaptic weight"; to continue with the analogy of the human brain, the synaptic weight is what our brain remembers (i.e., we are talking about the "data" stored in the human brain because of the learning process of everyone). After multiplying x_n by w_{nj}, all results are

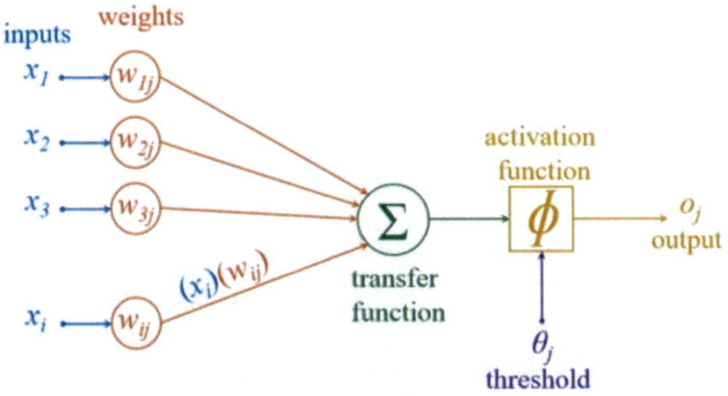

Fig. 13 Perceptron (artificial neuron) model. Adapted with permission from [21]. ©2020 IEEE

combined through the transfer function $\sum$ (which is essentially a simple sum, sometimes corrected by a number called "bias"). The result of these operations is then filtered by the "activation function" ϕ which produces the output o_j.

At this point the challenge is how to map each component of a neural network to the NAND architecture of a Flash memory and this is the subject of the following section.

3.1 Conventional 3D NAND

The translation of Fig. 13 in the NAND domain is depicted in Fig. 14. Before making a 1:1 comparison between Figs. 13 and 14 it is necessary to introduce another element to the discussion. To be more specific, input voltages V_{gi} are translated into "synaptic currents" I_{ij} by using the analog multiplier depicted in Fig. 15, which is made of two

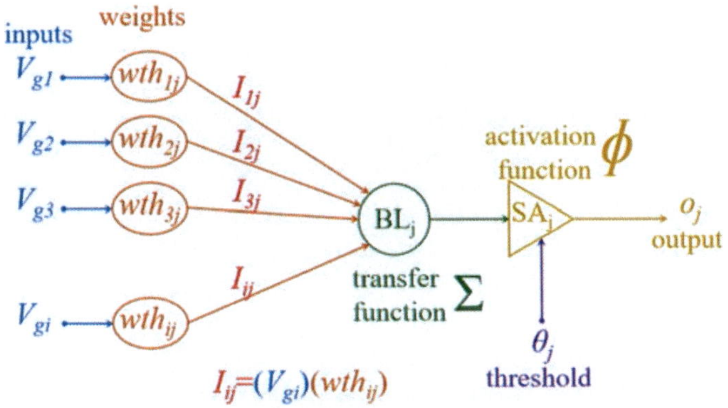

Fig. 14 NAND-based Perceptron model. Adapted with permission from [21]. ©2020 IEEE

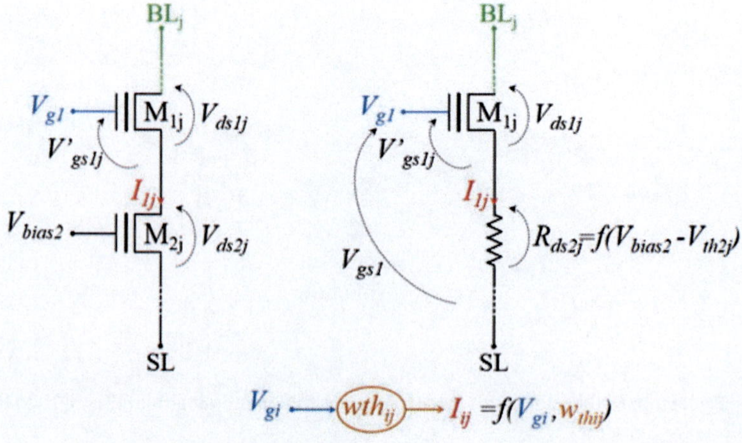

Fig. 15 Circuit schematic of Synaptic Cell (SC) with biases and currents indicated. The concept of the figure is based on the description provided in [21]

floating gate cells connected in a NAND fashion. For sake of simplicity, in the following we'll refer to the structure of Fig. 15 as a Synaptic Cell or SC.

BL and SL stand for BitLine and Source Line, respectively. FG transistors M_{2j} are intentionally biased in the triode region such that they can act as a resistor.

If we assume that the voltage applied to SL is ground, then we can write the following:

$$V'_{gs1j} = V_{gs1} - I_{1j} \cdot R_{ds2j} \tag{7}$$

For FG transistors of width W, length L, oxide capacitance C_{ox}, and threshold voltage V_{th}, the following equation holds true:

$$I_{1j} = \frac{\mu_n \cdot C_{ox}}{2} \cdot \frac{W}{L} \cdot \left(V'_{gs1j} - V_{th1j}\right)^2 \tag{8}$$

Therefore, by combining (7) and (8), it follows that

$$I_{1j} = \frac{\mu_n \cdot C_{ox}}{2} \cdot \frac{W}{L} \cdot \left(V_{gs1} - I_{1j} \cdot R_{ds2j} - V_{th1j}\right)^2 \tag{9}$$

As FG transistor M_{2j} is kept in the triode region by design (V_{bias2} and V_{th1j} properly set), its equivalent resistance value R_{ds2j} can be written as

$$R_{ds2j} = \frac{V_{ds2j}}{I_{1j}} \approx \frac{1}{k2 \cdot \left(V_{bias2} - V_{th2j}\right)} \tag{10}$$

By substituting (10) in (9) and applying a simple math we can write that

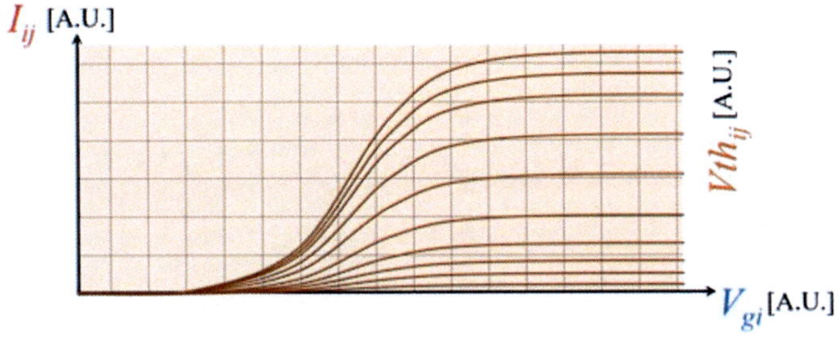

Fig. 16 Synaptic current I_{ij} as a function of V_{gi} and V_{thij}. Adapted with permission from [21]. ©2020 IEEE

$$I_{1j} \approx \frac{k_1 \cdot k_2}{4k_1} \cdot \left(V_{gs1} - V_{th1j}\right) \cdot \left(V_{bias2} - V_{th2j}\right) \tag{11}$$

which, in essence, is the transconductance G_m of the synaptic cell and can be written as

$$I_{1j} \approx G_m\left(V_{gs1}, V_{th2j}\right)|_{V_{bias2}, V_{th1j}} \tag{12}$$

The relationship between I_{ij} and V_{gi} as a function of the threshold voltage V_{thij} is shown in Fig. 16.

Multiple NAND strings sharing the same bitline can be used to form a single perceptron (Fig. 17); a 3D view of the corresponding schematic is given in Fig. 18.

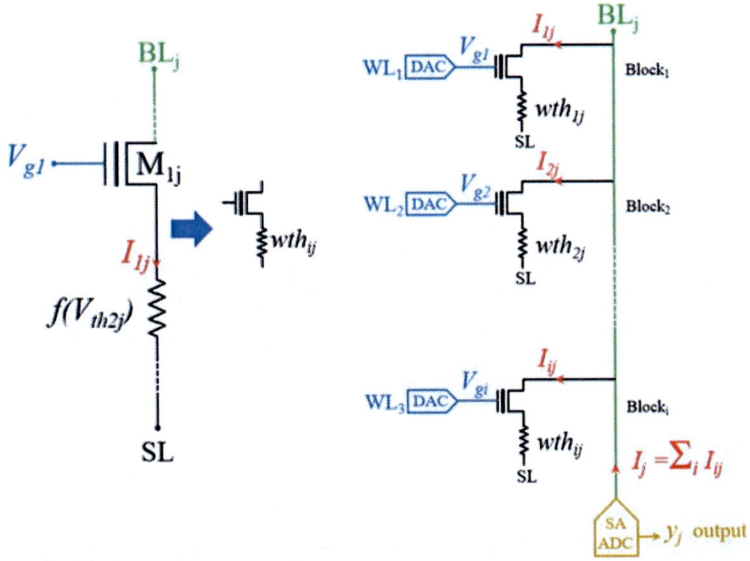

Fig. 17 Transistor-level description of one perceptron

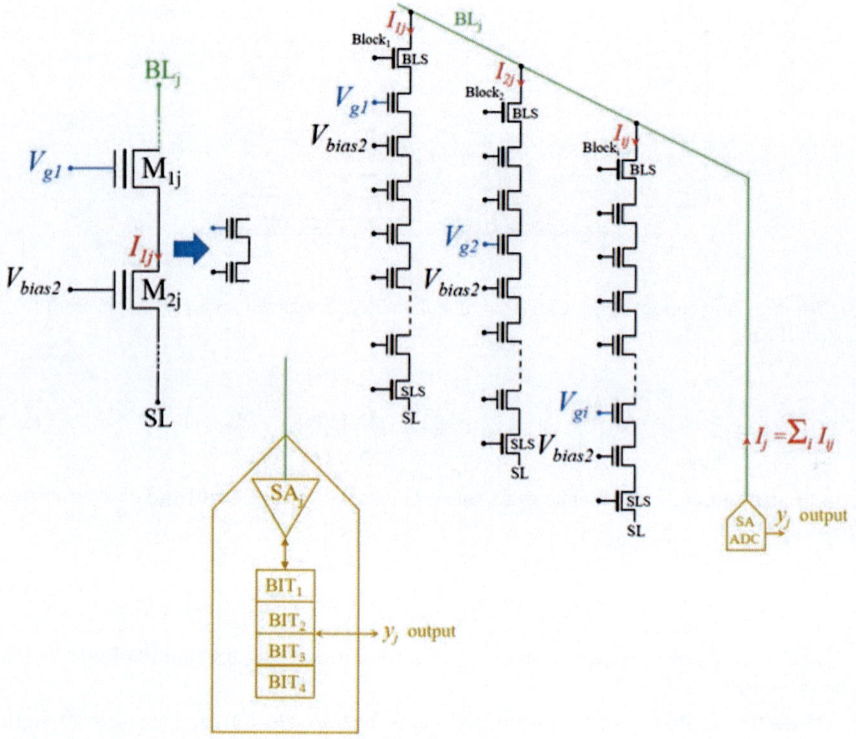

Fig. 18 3D schematic of one perceptron implemented with NAND Flash. The concept of the figure is based on the description provided in [21]

In a nutshell, all the currents flowing through the NAND strings are summed on the bitline thanks to Kirchoff's law and then converted into an output voltage by the Sense Amplifier SA. Please note that in a conventional NAND architecture, only one string is sensed per bitline (in other words, in a standard NAND, currents are not summed).

As we speak, NAND chips, given the number of available NAND strings, can implement more than 100 k perceptrons per chip, which is a remarkable result if compared to a standard implementation by means of logic gates.

3.1.1 Conventional-NAND-Based Single Layer Perceptron

Let's now get into the details of the implementation of the connections of multiple neurons by referring to, for the sake of simplicity of the drawings, a single layer neural network, as sketched in Fig. 19.

There are *i* neurons in the input layer and *j* neurons in the output layer. Input voltages V_{gi} are applied to multiple synapses, each one with different weights as

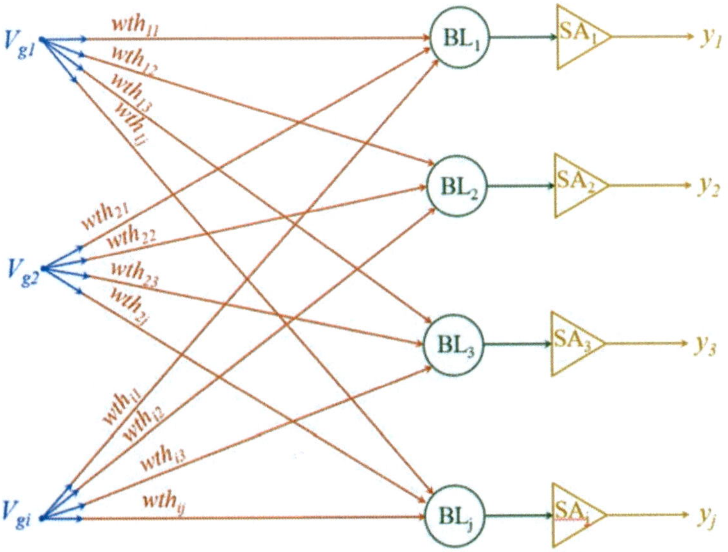

$$\text{weight matrix } W_{ij} = \begin{bmatrix} wth_{11} & wth_{12} & wth_{13} & \cdots & wth_{1j} \\ wth_{21} & wth_{22} & wth_{23} & \cdots & wth_{2j} \\ \vdots & \vdots & \vdots & & \vdots \\ wth_{i1} & wth_{i2} & wth_{i3} & & wth_{ij} \end{bmatrix}$$

Fig. 19 Representation of a NAND-based neural network with multiple neurons (top) and the corresponding weight matrix (bottom)

specified in the weight matrix. As usual, weights are computed offline during the training phase [22].

Figure 19 can be translated into a NAND architecture by using bitlines and wordlines as represented in Fig. 20. To be more specific, the input voltage V_g is applied to a single wordline such that it is applied in parallel to multiple NAND strings. As a result, each wordline implements a fully connected layer to a neuron of the previous layer of the neural network.

We can go even further and map the implementation of Fig. 20 to a 3D NAND, as shown in Fig. 21, where $Block_i$ corresponds to a physical 3D NAND block.

A bird's eye view of Fig. 21 is shown in Fig. 22, where all the basic elements, like bitlines, wordlines, NAND blocks, and sense amplifiers are shown.

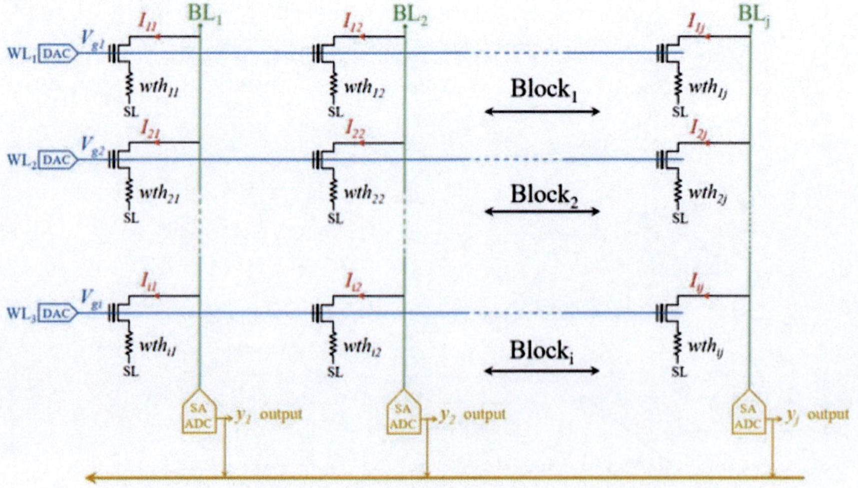

Fig. 20 Transistor level representation of Fig. 19. Adapted with permission from [21]. ©2020 IEEE

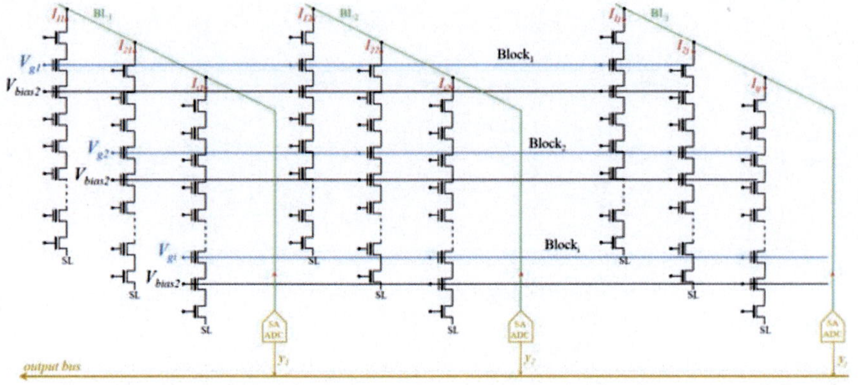

Fig. 21 3D NAND implementation of Fig. 20. The concept of the figure is based on the description provided in [21]

3.1.2 Conventional-NAND-Based Multi-Layer Perceptron

For real applications, a single layer perceptron is not good enough; neural networks with multiple layers are required to meet the target accuracy and precision [23]. To implement a feed-forward architecture, each layer, between input and output, is connected to another layer, as shown in Fig. 23. Layers sitting between input and output layers are called "hidden". To port Fig. 23 into the NAND domain, a multiple iteration approach is required. Indeed, it is worth recalling the fact that the output of each sense amplifier of a NAND device is digital in nature; therefore, it must be

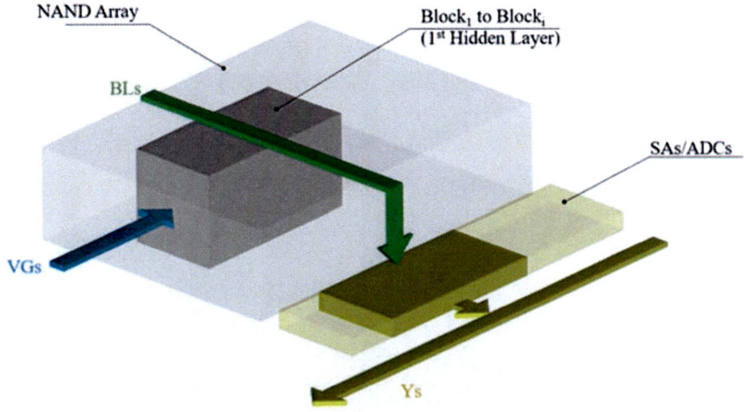

Fig. 22 3D NAND used to implement a single layer artificial neural network

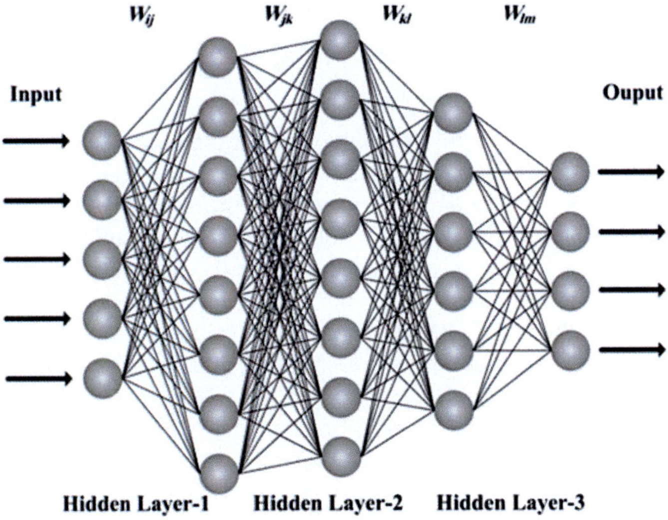

Fig. 23 Artificial neural network with three hidden layers

translated back into the analog domain before being applied to the wordlines of the following layer.

From a NAND perspective, Fig. 23 translates into Fig. 24. Multi-level NAND devices (i.e., NAND storing more than one digital bit into the same physical cell) contain digital-to-analog converters with fine granularity to reduce the width of the threshold voltage distributions as much as possible. These circuits can be considered as the enabling factor of the NAND-based multi-layer ANN. In Fig. 24 each gray box represents a hidden layer. To make the ANN work, in the first iteration layer 1

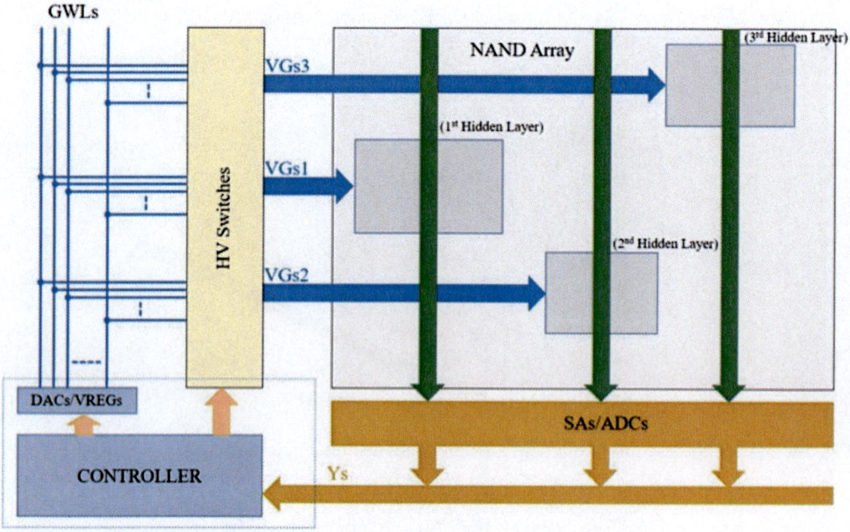

Fig. 24 NAND architecture to implement a multi-layer ANN

is red and then the corresponding (digital) output sitting inside the sense amplifier is sent to the block performing the activation function.

At this point we are almost ready to start the second phase but, before going there, it is necessary to translate back into the analog domain the output of the activation function. As already mentioned, this can happen thanks to the digital-to-analog converter normally used during both the reading and the programming phases of a standard multi-level NAND. Once the analog voltage is available, phase 2 can start, essentially repeating all the steps described for the first phase, but this time on layer 2.

Assuming a 4 bit/cell NAND (QLC), currents can be converted with a 4-bit precision as sketched in Fig. 25. The activation function is applied in the digital domain, by using a look-up-table (LUT); thanks to this approach, precision can be increased up to 10 bits (this oversampling is necessary to have a good implementation of the activation function). Of course, the output of the activation function has then to be translated back into the analog domain to be applied to the wordlines of the next ANN layer inside the Flash memory array.

3.2 3D NAND with Source-Line Read

The VbM multiplication can also be implemented by using a 3D NAND memory where the read voltage V_{READ} is applied to the Source Line; a schematic representation of this 3D array is given in Fig. 26 [24].

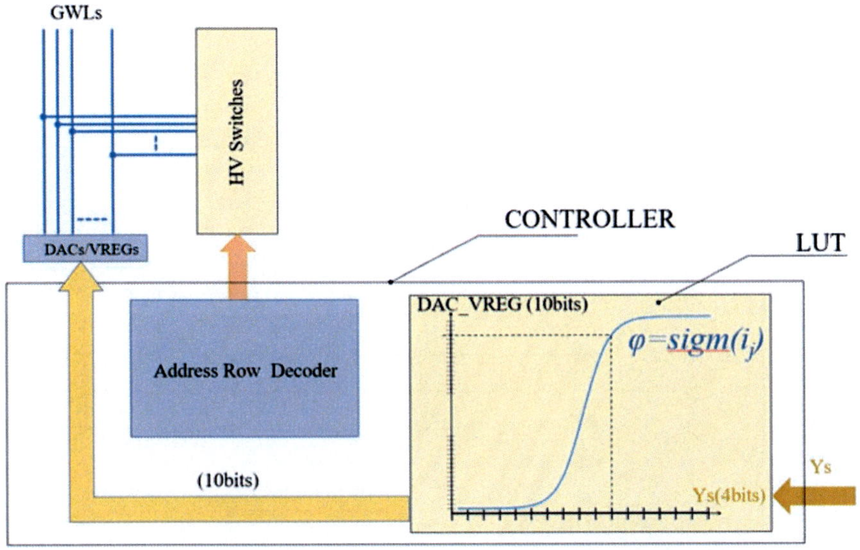

Fig. 25 Sigmoid activation function with QLC NAND. Adapted with permission from [21]. ©2020 IEEE

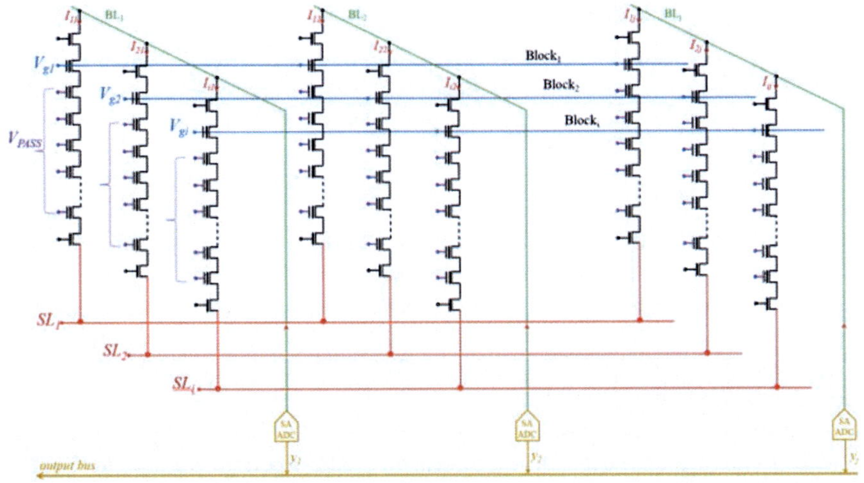

Fig. 26 3D NAND memory array with Source-Line read. The concept of the figure is based on the description provided in [24]

Let's now review the biasing scheme when the VbM multiplication needs to take place (Fig. 27). First, bitlines (BLs) are set at GND (ground), while the read voltage V_{READ} is applied to source lines (SLs). Voltage V_{PASS} is then used for all the wordlines belonging to unselected layers. Like in a conventional NAND architecture, V_{PASS}

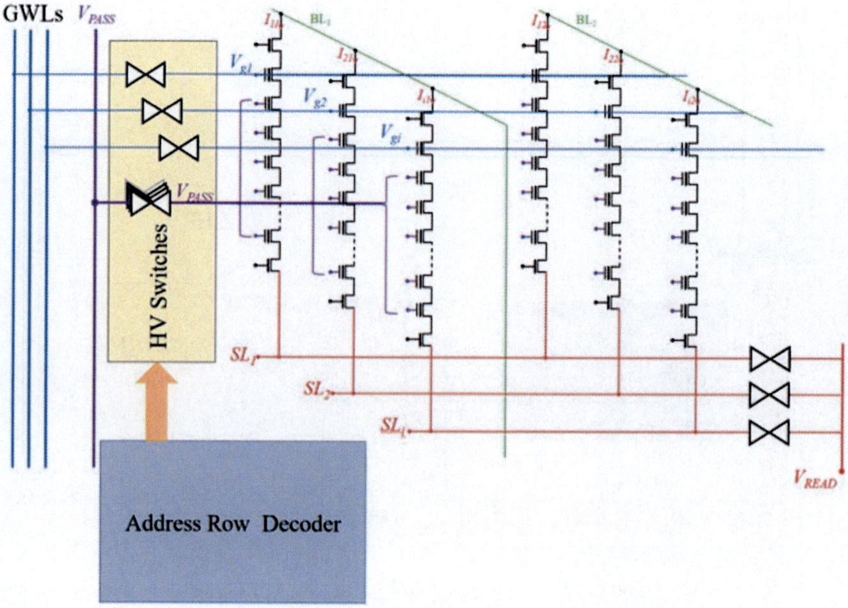

Fig. 27 Biasing scheme of Fig. 26 when the VbM multiplication takes place. The concept of the figure is based on the description provided in [24]

needs to be high enough to make unselected cells ON independently from their threshold voltage. Within the selected 3D layer, GND and V_{SEL} are used; which one to apply depends upon the input pattern.

As in the previous architecture, currents are summed on the bitline node, thus implementing a weighted sum per column, as required by the VbM multiplication. Analog-to-digital conversion is performed by the ADC converters placed at the end of the bitlines.

It is worth highlighting the importance of choosing the right biasing conditions as they directly influence the accuracy of the neural network. As already mentioned, V_{PASS} needs to make every cell conductive; in [24], V_{PASS} is set to 8 V, while V_{SEL} is a value between 1 and 2.5 V.

Another aspect to be considered when designing the VbM multiplication is the read latency, i.e., the time it takes to perform the analog operations within the NAND array. Figure 28 sketches the main parasitic elements that need to be considered when looking at electrical performances. Indeed, parasitics, resistors and capacitors, end up slowing down the propagation of the signals. Therefore, the overall performance

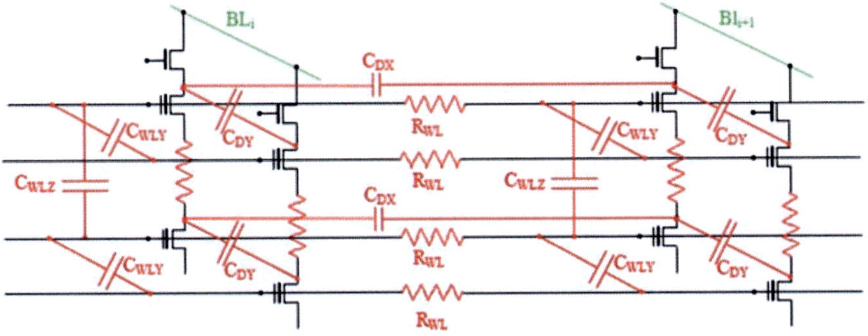

Fig. 28 Parasitics of a 3D NAND array. The concept of the figure is based on the description provided in [24]

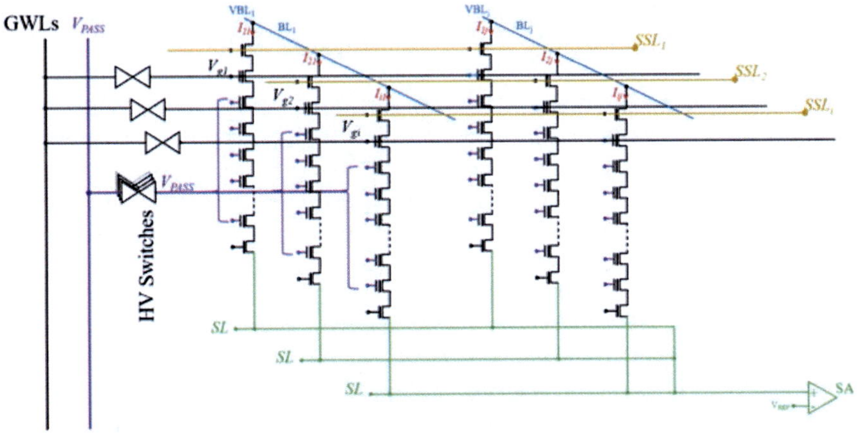

Fig. 29 Building block of a 3D NAND nvCIM. The concept of the figure is based on the description provided in [25]

of a NAND chip is dictated by both the VbM parallelism and the speed of a single VbM multiplication, which is directly coupled to the physical implementation of the 3D memory array itself.

3.3 3D NAND with Independent Source Lines

Another architecture for performing the VbM multiplication has been proposed in [25], namely the nvCIM architecture, which stands for non-volatile Computing-In-Memory; this is again a 3D NAND based architecture, but this time with independent Source Lines.

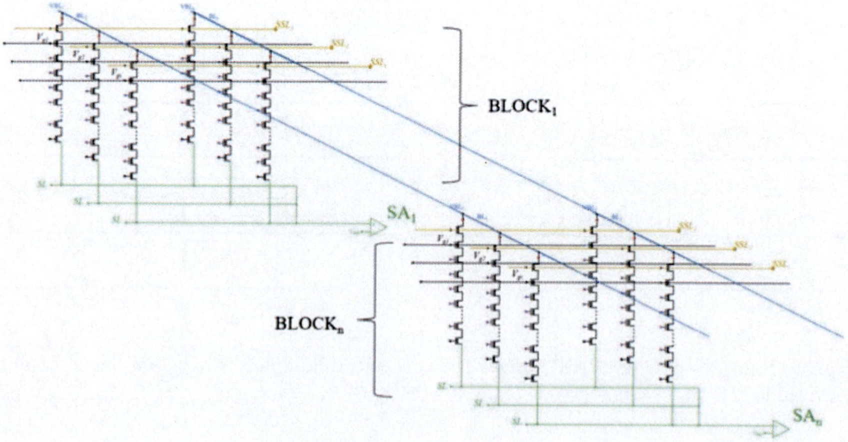

Fig. 30 3D NAND nvCIM. The concept of the figure is based on the description provided in [25]

The schematic of nvCIM is sketched in Fig. 29. Bitline voltages VBL_i are the inputs to the VbM multiplication. Only a single 3D layer at a time is selected by applying the same voltage Vg to every Vgi. The current flowing through the NAND string is, as a matter of fact, the product of the input voltage by the conductance of the selected memory cell. Currents are then summed at the source terminal of the block and converted to a digital value by the sense amplifier SA. Please note that the sense amplifier sitting at the end of the source line is the only additional element compared to a conventional NAND architecture. As sketched in Fig. 30, one SA per NAND block is required, which means that each block has its own source line.

Of course, because multiple blocks are available within a NAND die, multiple VbM multiplications can be computed in parallel.

As we have seen in this section, there are multiple ways for leveraging a NAND array when dealing with the VbM multiplication. This is especially true if the NAND architecture is coupled with a multi-level approach, i.e., the ability of storing more bits inside the same physical cell; this aspect is key to enable the implementation of low power and, at the same time, precise artificial neural networks. All the works mentioned above testify the effort of the scientific community to make neural networks the next killing application for NAND Flash memories.

References

1. C. Mead, *Analog VLSI and Neural Systems* (Addison-Wesley, Reading, MA, USA, 1989)
2. G. Indiveri et al., Neuromorphic silicon neuron circuits. Frontiers Neurosci. **5**(73), 1–23 (2011)
3. K. Likharev, CrossNets: neuromorphic hybrid CMOS/nanoelectronic networks. Sci. Adv. Mater. **3**, 322–331 (2011)
4. J. Hasler, H. Marr, Finding a roadmap to achieve large neuromorphic hardware systems. Front. Neurosci. **7**. Article no. 118

5. L. Ceze et al., Nanoelectronic neurocomputing: status and prospects, in *Proceedings of the DRC, Newark*, DE, USA, June 2016, pp. 1–2
6. D.B. Strukov, H. Kohlstedt, Resistive switching phenomena in thin films: materials, devices, and applications. MRS Bull. **37**(2), 108–114 (2012)
7. S. Raoux, D. Ielmini, M. Wuttig, I. Karpov, Phase change materials. Mater. Res. Soc. Bull. **37**(2), 118–123 (2012)
8. W. Lu, D.S. Jeong, M. Kozicki, R. Waser, Electrochemical metallization cells—blending nanoionics into nanoelectronics? MRS Bull. **37**(2), 124–130 (2012)
9. J.J. Yang, I.H. Inoue, T. Mikolajick, C.S. Hwang, Metal oxide memories based on thermochemical and valence change mechanisms. MRS Bull. **37**(2), 131–137 (2012)
10. E.Y. Tsymbal, A. Gruverman, V. Garcia, M. Bibes, A. Barthélémy, Ferroelectric and multiferroic tunnel junctions. MRS Bull. **37**(2), 138–143 (2012)
11. S. Park et al., RRAM-based synapse for neuromorphic system with pattern recognition function, in *IEDM Technical Digest*, Dec 2012, pp. 10.2.1–10.2.4
12. Y. Nishitani, Y. Kaneko, M. Ueda, Supervised learning using spike-timing-dependent plasticity of memristive synapses. IEEE Trans. Neural Netw. Learn. Syst. **26**(12), 2999–3008 (2015)
13. M. Prezioso, F. Merrikh-Bayat, B.D. Hoskins, G.C. Adam, K.K. Likharev, D.B. Strukov, Training and operation of an integrated neuromorphic network based on metal-oxide memristors. Nature **521**, 61–64 (2015)
14. S. Kim et al., NVM neuromorphic core with 64k-cell (256-by-256) phase change memory synaptic array with on-chip neuron circuits for continuous in-situ learning, in *IEDM Technical Digest*, Dec 2015, pp. 443–446
15. X. Guo et al., Fast, energy-efficient, robust, and reproducible mixed-signal neuromorphic classifier based on embedded NOR flash memory technology, in *2017 IEEE International Electron Devices Meeting (IEDM)* (2017), pp. 6.5.1–6.5.4. https://doi.org/10.1109/IEDM.2017.8268341
16. F. Merrikh-Bayat, X. Guo, M. Klachko, M. Prezioso, K.K. Likharev, D.B. Strukov, High-performance mixed-signal neurocomputing with nanoscale floating-gate memory cell arrays. IEEE Trans. Neural Netw. Learn. Syst. **29**(10), 4782–4790 (2018). https://doi.org/10.1109/TNNLS.2017.2778940
17. SST Inc., Superflash technology overview. www.sst.com/technology/sst-superflash-technology. Accessed 7 Feb 2017
18. C.R. Schlottmann, P.E. Hasler, A highly dense, low power, programmable analog vector-matrix multiplier: the FPAA implementation. IEEE J. Emerg. Sel. Topics Circuits Syst. **1**(3), 403–411 (2011)
19. Y.J. Park, H.T. Kwon, B. Kim, W.J. Lee, D.H. Wee, H.S. Choi, B.G. Park, J.H. Lee, Y. Kim, 3-D stacked synapse array based on charge-trap flash memory for implementation of deep neural networks. IEEE Trans. Electron. Devices **66**, 420–427 (2019)
20. Y.J. Park et al., 3-D stacked synapse array based on charge-trap flash memory for implementation of deep neural networks. IEEE Trans. Electron Devices **66**(1), 420–427 (2019). https://doi.org/10.1109/TED.2018.2881972
21. U. Minucci, L.D. Santis, T. Vali, F. Irrera, A neural network implemented on NAND memory, in *2020 IEEE International Memory Workshop (IMW)*, (2020), pp. 1–4. https://doi.org/10.1109/IMW48823.2020.9108138
22. S. Yu, Neuro-inspired computing with emerging nonvolatile memory. Proc. IEEE **106**(2) (2018)
23. G.W. Burr et al., Neuromorphic computing using non-volatile memory. Adv. Phys. X **2**(1), 89–124 (2017)
24. P. Wang et al., Three-dimensional nand flash for vector–matrix multiplication. IEEE Trans. Very Large Scale Integr. (VLSI) Syst. **27**(4), 988–991 (2019). https://doi.org/10.1109/TVLSI.2018.2882194
25. D. Ielmini, H.P. Wong, In-memory computing with resistive switching devices. Nat. Electron. **1**, 333–343 (2018)
26. Y. LeCun, C. Cortes, C. Burges, MNIST handwritten digit database, http://yann.lecun.com/exdb/mnist/

Machine Learning for 3D NAND Flash and Solid State Drives Reliability/Performance Optimization

Cristian Zambelli, Rino Micheloni, and P. Olivo

Abstract Solid State Drives (SSDs) are the storage backbone of many applications ranging from consumer electronics up to exa-scaled data centers (Zuolo in Proc IEEE 105:1589–1608, 2017). As such, the performances and reliability of SSDs should be improved to reduce the Total Cost of Ownership of the system hosting the drive. This activity requires a careful evaluation of those figures exposed by the integrated storage medium, namely the 3D NAND Flash memory (Micheloni in Proc IEEE 105:1634–1649, 2017). However, as this non-volatile memory technology scales and evolves, there is an ever-growing number of tuning knobs to leverage for such assessments and optimizations. Machine learning has emerged as a viable solution in many stages of this process. After introducing the reliability issues of the 3D NAND Flash memories, this chapter shows both supervised and un-supervised machine learning techniques used to identify homogeneous areas inside the Flash array, thus enabling an optimization of the storage system performance through Error Correction Code (ECC) fine-tuning. Moreover, this chapter deals with algorithms and techniques for a pro-active reliability management of SSDs utilized in a smart monitoring system being storage media agnostic. The last section of the chapter discusses the next challenge for machine learning in the context of SSDs for enterprise applications, namely the Computational Storage (CS) paradigm based on the co-integration of Field Programable Gate Arrays (FPGAs) and storage devices.

C. Zambelli (✉) · R. Micheloni · P. Olivo
Dipartimento di Ingegneria, Università degli Studi di Ferrara, Ferrara, Italy
e-mail: cristian.zambelli@unife.it

R. Micheloni
e-mail: rino.micheloni@unife.it

P. Olivo
e-mail: piero.olivo@unife.it

R. Micheloni and C. Zambelli (eds.), *Machine Learning and Non-volatile Memories*,
https://doi.org/10.1007/978-3-031-03841-9_7

1 Introduction

Solid State Drives (SSDs) are the storage backbone of many applications ranging from consumer electronics up to exa-scaled data centers [1]. The performance and the reliability figures of an SSD should be improved to reduce the Total Cost of Ownership of the system hosting the drive. This requires a careful evaluation of those figures exposed by the integrated storage medium, namely the 3D NAND Flash memory [2]. However, as this non-volatile memory technology scales and evolves under many viewpoints, there is an ever-growing number of tuning knobs to leverage for such assessments and optimizations. Machine learning emerged as a viable solution in many stages of this process.

This chapter describes how some of the machine learning techniques discussed in this book can be exploited either for the reliability characterization of the 3D NAND Flash technology as a support for the development of system-level approaches to be embodied in SSD controllers or as monitoring tools agnostic of the storage medium. The Sect. 2 briefly introduces the reliability issues of the 3D NAND Flash memories evidencing how the manufacturing process poses challenges in their assessment due to the large intrinsic variability of the technology. The Sect. 3 shows different supervised and un-supervised machine learning techniques that are exploited to find homogeneous reliability spots in the memory to enhance the storage system performance through Error Correction Codes (ECC) fine-tuning. In addition, we will discuss deep learning methods to be coupled with the ECC engine embedded inside the SSD and methods for variability-aware storage. In Sect. 4, there will be a presentation of algorithms and techniques for pro-active reliability predictions of SSDs implemented in a smart monitoring system being storage media agnostic, therefore not directly addressing 3D NAND Flash. Finally, Sect. 5 discusses the foundation of the machine learning frontiers for SSDs in enterprise applications, namely the Computational Storage (CS) paradigm based on the co-integration of Field Programable Gate Arrays (FPGAs) and storage devices.

2 Reliability and Variability Issues of 3D NAND Flash

In the last decade, different 3D NAND Flash architectures [3–6] have been proposed as the storage medium for multi terabits SSD applications. Despite the peculiarities of each of them in terms of materials, cells' array topological organization, and peripheral circuitry placement strategies, there is always an important feature to assess above all, namely their reliability. It is worth to say that the 3D NAND Flash technology inherits all the issues that have been investigated in the past for its planar counterpart, so the struggle to achieve a significantly high endurance, data retention, and read disturb immunity to be qualified as a valid mass storage technology is still there [7, 8]. In addition, since 3D devices (i.e., transistors) behavior must account for some specific physical principles, there is to report either an exacerbation of issues which were partially solved in the past (like the cross-temperature phenomenon [9]) or radically new reliability threats (i.e., like the Temporary Read Errors) investigated at product-level as in [10].

The paradigm changes in the manufacturing process (stepping from the litho-intensive process, typical of planar Flash devices, to an etch-intensive one) produces significant deviations from a perfect cylindrical structure that should represent the 3D NAND Flash memory string. Diameter variations (unavoidable in all architectures due to the high aspect ratio required for creating hundreds of stacked layers) coupled with mechanical stress issues that can cause bows or twists (see Fig. 1), may easily alter the memory characteristics. In [11], there is a report of how those variations in a 3D NAND Flash block change the reaction time to the voltage stimuli applied during the memory operation due to the different resistance and capacitance exhibited by the different layers. It is not difficult to speculate how this would affect reliability. Since each layer will behave as an independent entity, there is an intrinsic complexity in controlling the outcome or read, program, and erase operations, resulting in long term degradation of the reliability metrics like endurance or retention. This phenomenon is generally referred as layer-to-layer variability [12]. Storage systems like SSDs would perceive this by experiencing a large difference in the number of corrupted bits (i.e., fail bits) after a specific operation performed in different topological areas of a 3D NAND Flash chip. As an example, let us assume that the memory in the SSD has undergone endurance stress (i.e., repeated program and block erase operations): besides the huge fail-bit count difference expected between different page types if the memory adopts a multi-bit per cell paradigm (Triple Level Cell—TLC or Quadruple Level Cell—QLC), there are topological regions more prone to errors than others. The reliability in that region of the 3D NAND Flash block may even exceed the correction capabilities of Error Correction Codes (ECC) engines integrated in the SSD controller, thus resulting in potential data corruption [12]. This generic behavior holds true for other issues of the 3D NAND Flash technology that were not deeply investigated in planar technology, such as the cross-temperature [9].

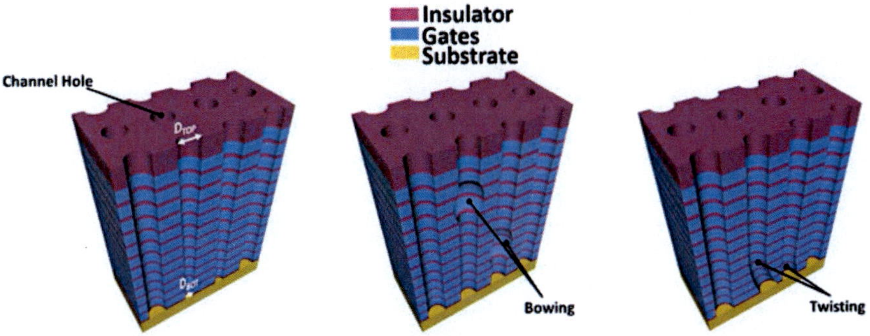

Fig. 1 Example of process issues in the 3D NAND Flash fabrication that will affect layer-to-layer variability and reliability. Reproduced with permission from [12]. © 2019 IEEE

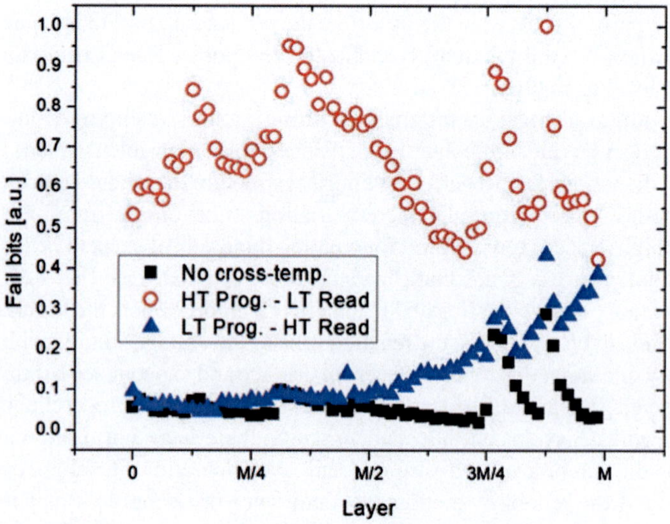

Fig. 2 Layers sensitivity to cross-temperature issues shown by the normalized fail bits count (FBC) to the device worst-case in 3D NAND Flash memories. Reproduced with permission from [12]. © 2019 IEEE

Figure 2 shows that a large layer-to-layer variation makes almost impossible to identify a specific trend as a function of the layer position and, therefore, it is very hard to tackle the reliability loss due to the cross-temperature effect by simple algorithms. As a result, 3D NAND Flash vendors started to progressively introduce a set of tweaking knobs directly in the memory chip that storage system designers may leverage to countermeasure the variability and reliability issues induced by manufacturing process criticalities [13, 14]. Among them, we report the possibility to change the read voltages to track the shift of threshold voltage distributions induced by endurance, retention or read disturb (known as the Moving Read Reference [15]), modify the program, and erase strategies, and so on. All these aids can be exploited by the Flash Signal Processing algorithms executed by the SSD controller, through its custom hardware accelerators and associated firmware. However, the development of the abovementioned algorithms requires a time-consuming characterization of 3D NAND silicon that generates a huge amount of data to analyze; the problem is that, especially with hundreds of layers, sometimes it is very difficult to come up with easy-to-implement algorithms because of the multitude of options to consider. Machine learning has then emerged as a possible relief in this context.

3 Machine Learning Techniques for 3D NAND Flash

The point in time where storage designers started to look at machine learning techniques applied to 3D NAND Flash memories was when the "judge-by-eye" approach started to fail in the reliability management algorithms development process. The many available ECC strategies [16] implemented in the SSD controller as well as advanced read oversampling correction schemes [17] could not find a perfect fit for every corner of the memory. As an example, solutions that are a best fit for endurance stress would turn sub-optimal for retention stress. Even in the last nodes of planar NAND Flash technology (i.e., mid-1X) the designers had to face, although with a reduced effort, a similar challenge. Machine learning was then reported in this context in seminal literature works like those in [18–20]. With the advent of 3D NAND Flash technology and with the momentum gained by the artificial intelligence also in the storage community, there is a surge of interest in exploiting techniques from unsupervised clustering up to neural networks for characterization data analysis and prediction. In this section, we report few of them at work in some specific 3D NAND Flash test conditions and with slightly different implementation goals (i.e., real-time predictions or off-line models' development).

3.1 Unsupervised Learning (Clustering) for ECC Optimization

The data clustering algorithm is a must-have in the unsupervised machine learning toolbox to find homogeneous spots (i.e., clusters) in an ensemble of data representing an observable phenomenology. The work in [21] applied this concept to the evaluation of the endurance stress reliability in a TLC 3D NAND Flash off-the-shelf product. The goal of the work was to identify memory regions in the chip that behave similarly when tested with the same stress to develop a proper ECC strategy for each of those regions. Indeed, balancing the implementation cost with the reliability/performance trade-off exercised by the memory is an important task for advanced codes like the Low-Density Parity-Check (LDPC) [22]. In Flash-based SSDs there is a demand for very high code rates (i.e., the ratio between the user data and the codeword data including parity bits), meaning a small overhead in parity bits but offering less correction capability. On the other hand, while a low code rate could achieve a significantly good correction capacity it imposes a severe user capacity waste that increases the storage costs in turn [23].

The dataset used in the development of a clustering strategy according to [21] consisted in readouts of all the topological memory locations. The data are the fail bit counts retrieved on 4 KB chunks, which is both the minimum unit read during tests and the data size on which the ECC is expected to work in all the fail bits correction activities. The characterization was performed on more than 800 different memory blocks spread on different TLC 3D NAND Flash chips and manufacturing lots to account for the process-induced variability [12]. Analyzing such a large statistical population obtained after the endurance stress allowed some speculations on the Bit Error Rate (BER) sustainable by the ECC in one of the worst reliability corners for the memory. The dataset was then used as input to a *k-means* clustering algorithm [24], whose goal is to partition the dataset in *k* clusters of homogeneity. The algorithm applied to endurance stress evidenced that unsupervised learning could find specific regions of the memory with a similar BER, although not readily usable by an SSD controller. It was found that the clustering procedure needs to be enhanced through a semi-supervised approach in which some additional rules are considered, namely the separation of different TLC pages and the grouping of all the data chunks composing a TLC page into a single cluster. This approach is also defined as constrained clustering, whose result is shown in Fig. 3. Six different regions have been identified. A system designer can then decide to allocate a proper ECC code (e.g., an LDPC) per region. Without clustering, SSD controller vendors should develop correction algorithms by considering the worst BER of the whole 3D NAND Flash chip, but such an assumption would be extremely conservative and sacrifices a lot of user space in the SSD. When data clustering is used instead, it is found that the ECC code rate can be tailored per TLC page type and per cluster, materializing in 24% gain on the SSD addressable space.

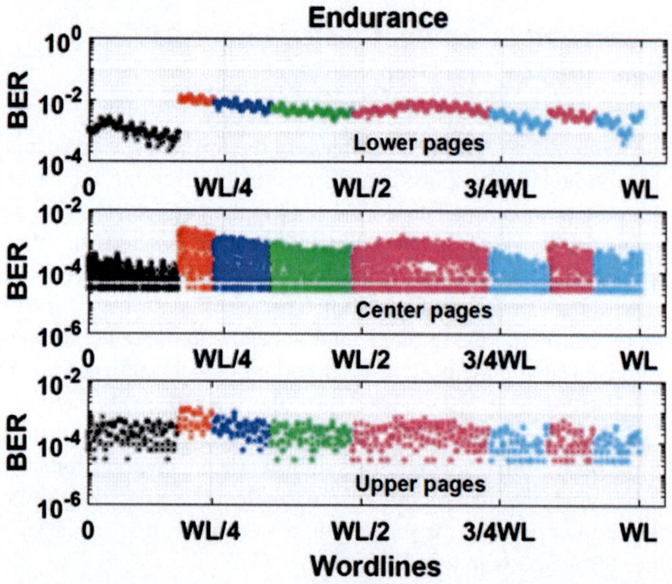

Fig. 3 Constrained clustering results on 3D-NAND flash. Six different clusters have been identified in the 3D NAND Flash endurance characterization dataset. Reproduced with permission from [21]. © 2017 IEEE

3.2 Supervised Learning for Pre-emptive Endurance Prediction

The previous section discussed an off-line technique for LDPC optimization in the context of the endurance failures. However, we must remind that the outcome of that study was possible thanks to the access to a huge test population, which sometimes is not feasible for SSD controller designers. In this context, the study in [25] proposes to create a machine learning algorithm deployable in the firmware of SSDs that pre-emptively predicts the endurance failures of the Flash memory by tracking the BER worst-case regions during the lifetime of the drive, therefore achieving an on-line reliability management methodology. This approach is based on a supervised learning technique (chapter "Introduction to Machine Learning"), namely the Support Vector Machine (SVM) [26]. The claim of [25] is to reduce the uncorrectable error rate due to endurance failures by spreading the cycling across all memory blocks in a better way than a simple wear-leveling algorithm [27]. A similar claim has been proposed in [28], although other supervised learning techniques like Random Forest (RF) and Logistic Regression (LR) have been implemented along with the SVM. The ease of implementation of these machine learning techniques in any of the statistically capable data elaboration software and the consequent hardware porting on any computing unit such as an embedded processor can envision the endurance prediction for each block as a background task for the SSD, introducing thus minimal latency overhead. All the research in the context of supervised learning mainly focuses on the features of the dataset. Working with 3D NAND Flash characterization data usually leads to a strongly unbalanced dataset. Indeed, testing a million of codewords in which only ten are failing due to some criteria will provide altered accuracy metrics of any statistical model. The models could have a 99.9% prediction accuracy since it predicts only the passing codewords in an endurance test. As pointed out in [25], it is important that models like SVM, Random Forest, etc., performs well also in sensitivity and specificity metrics rather than only accuracy. The sensitivity is indicated as a true positive rate (TPR) metric, whereas the specificity is the true negative rate (TNR) metric. Those values can be calculated starting either from the proportion of the codewords predicted to pass that effectively pass a test (TPR) or from the ratio between failing codewords effectively predicted to fail. If we take the average of the TPR and TNR metrics we can achieve a fair definition of the model accuracy. To balance the sensitivity and the specificity there are several statistical techniques mostly based on resampling [29] that makes possible to work on a balanced dataset.

An example of supervised learning application flow is reported in Fig. 4 from the study in [28] related to planar NAND Flash technology. Here, machine learning is applied to model memory cells suffering from Program Disturb (PD) due to endurance stress. The idea is to predict those cells at a specific number of Program/Erase cycles by using the knowledge base from cells screened during former endurance stress (i.e., prior Program/Erase cycles). Following a well-established procedure, 70% of cells are used for training, and 30% is kept for model evaluation. Three algorithms (LR, RF, and SVM) were evaluated reaching an initial accuracy

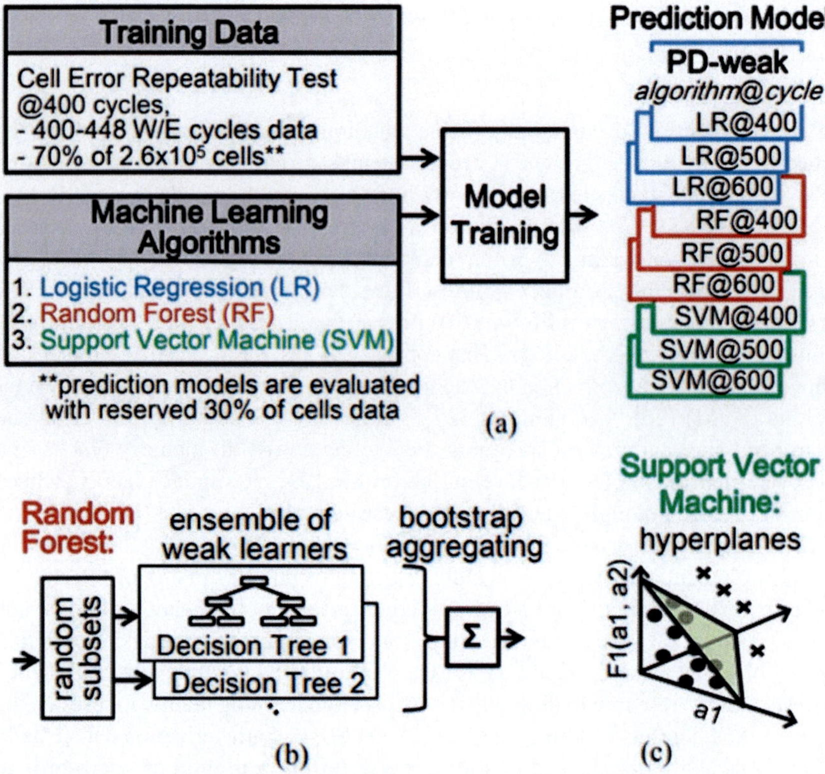

Fig. 4 **a** Predict future PD-weak cells after further W/E cycling by machine learning. **b** Random forest algorithm, **c** support vector machine algorithm. Reproduced with permission from [28]. © 2016 IEEE

of 40%. Despite this result may seem discouraging, the study in [25] performed on a similar technology shown that by applying both balanced sampling techniques and including additional characterization metrics (e.g., program time, erase time, etc.) it was possible to reach a prediction accuracy greater than 99% and a specificity/sensitivity between 98 and 99%. We can speculate that the modeling flow and the results obtained with the planar technology can be reproduced with 3D NAND Flash technology, although the higher intrinsic variability of the latter is a threat.

3.3 *Artificial Neural Networks Applied to Adaptive-ECC*

The 3D NAND Flash technology is not unique; different manufacturers use different 3D memory architectures and different materials (i.e., charge-trap or floating-gate) in the storage layer [3, 6]. This originates differences in the error characteristics related to some peculiar phenomena like retention loss or read disturb. Nowadays, having to deal with multiple 3D NAND technologies at the same time is common inside data centers as they host storage platforms from different vendors. Therefore, a wide variety of reliability challenges need to be addressed inside storage facilities. For many 2D Flash generations, improving the error correction scheme has been the solution to reliability problems, but this basic approach ran out of gas in 3D, especially in high availability and responsivity scenarios. In [30], the authors proposed to solve this complex issue by developing an Artificial Neural Network (ANN) engine coupled with the ECC integrated in the storage controller. This solution relies on LDPC codes and adaptively correct errors according to a trained error model of the memories used for storage. Using an ANN rather than a supervised learning model like those described in the previous sections of this chapter has different advantages in terms of prediction accuracy, speed and area overhead as disclosed in [30]. An important factor to consider when ANN are to be considered for deployment in storage systems is the test cost advantage. In the traditional memory test flow, the correlations between different topological parameters (e.g., wordline location, block number, die, etc.) and error characteristics can be captured by extensive tests whose results are recorded in huge look-up tables that increase in turn the storage costs [31]. With ANN, there is an initial price to pay for the hardware integration in the storage controller dedicated to the real-time inference of the neural network and the cost of the external infrastructure needed to train the network (generally GPUs), but when the network is deployed it can be used for any 3D NAND Flash technology that populates the storage (chapter "Introduction to Machine Learning"). Figure 5 shows the structure of the proposed ANN [30]. During the training phase, the error characteristics of the memory under test consists of a 4 MB sized random sample (i.e., 4 MB of memory cells) taken across the entire topological population of the memory (a 20 KB batch size is considered in this proposal). In the first training batch, the ANN is fed through the input neurons with parameters $x_1, \ldots, x_K$, where the k parameters can be any significant information to learn (e.g., V_{TH} state of the memory cells, TLC page type, wordline and layer index, etc.). A softmax (i.e., a mathematical function based on the logistic curve) is used as activation function for the output of the network, which is labelled as y_1 and y_2 for the "not error" and "error" probability. The synaptic weights of the network are updated by a backpropagation algorithm that minimizes the average loss function calculated on separate training batches as indicated in the Fig. 5. A total of 50 training epochs is exploited in the process. Once the ANN is trained, it is then ready to be deployed in the storage controller for the inference. The results claimed in [30] shows a 3D NAND Flash lifetime extension up to 9.1 times in some working conditions like data retention, proving the importance of neural networks in storage reliability prediction and management.

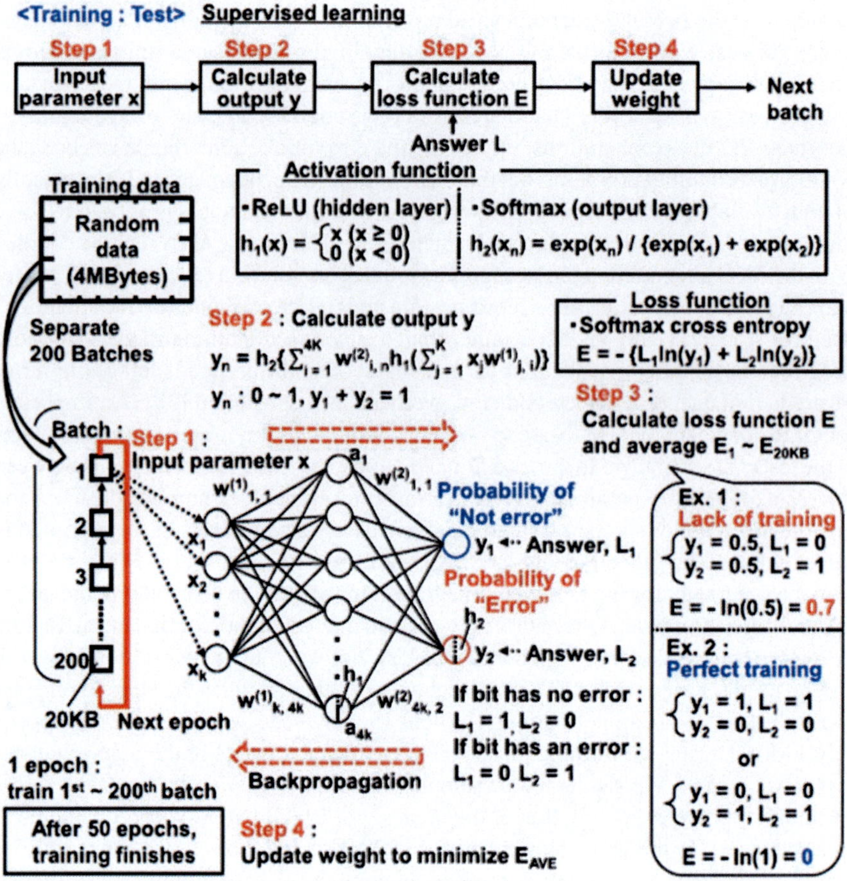

Fig. 5 Training of the proposed ANN (left). The proposed storage controller implements only the inference of ANN to correct errors (right). Reproduced with permission from [30]. © 2019 IEEE

3.4 Artificial Neural Networks for Variability Modeling

The errors behavior in 3D NAND Flash, as already highlighted in this chapter, is principally ascribed to the large technological variability of the manufacturing process. Several limitations on the 3D structure of the memory devices and on the materials integration [32] require the use of machine learning methods to accurately model their reflection at higher abstraction levels. This step relies on the approaches described in the previous sections. However, the system-level implementation complexity of those methodologies can be decreased if there is the possibility to apply machine learning methods to predict the technological variability sources since the production phase of the memory chips. From several works in literature [33–35], we know that those sources can be predicted by a simultaneous analysis of the fluctuations in physical process entities like poly-silicon channel defects, non-ideal taper angle of the channel, poor channel critical dimension control, etc. State-of-the-art modeling solutions are represented by extensive TCAD simulation campaigns that are both computationally and resource expensive, therefore becoming prohibitive for mass storage products analysis. In [36, 37], the authors proposed a new machine learning approach that can be applied in pre-production steps for 3D NAND Flash devices that achieves the same accuracy of TCAD simulations with a prediction error rate lower than 1% and a promised computational cost reduction up to 80%. The approach is based on a Multiple Input Multiple Output (MIMO) ANN whose input sources are divided in two kinds: variability sources input data and constant parameter input data. The variability sources input data correlates the variation magnitude of a specific source like defects, taper angle, and so on with the 3D NAND Flash device characteristics, whereas the constant parameter input data are values defining the electrical structure of the device (e.g., nitride thickness of the storage layer, filler oxide thickness, etc.). An ANN with three hidden layers featuring 100 neurons per layer is then trained based on this network topology. A ReLU activation function per layer (chapter "Neural Networks and Deep Learning Fundamentals") is considered to prevent issues in the solution search for the model. The output of the neural network is the predicted threshold voltage variation of the 3D NAND Flash cells (σV_{th}), the turn-on current variation of the Flash string (σI_{on}), the sub-threshold slope variation of the cells (σSS), and the trans-conductance variation (σg_m) which is important to estimate wear-out effects (this parameter is important to evaluate endurance stress as an example). Figure 6 shows the outcome of the network in retrieving effect the impact of the variations as function of the number of layers of 3D NAND flash memory. The model is tested to show the impact of the taper angle of the channel (related to the number of layers integrated) on the average value and on the standard deviation of a distribution of 3D NAND Flash cells evaluated in terms of electrical characteristic like those defined by the output neurons of the ANN. The results are compared with state-of-the-art TCAD simulations and are proven to be correct.

3.5 Recurrent Neural Networks for ECC Soft-Decision Speed Up

The 3D NAND flash memory suffers from a multitude of reliability issues that can be interpreted, paraphrasing the telecommunication terminology, as non-stationary noises in a communication channel that are difficult to be predicted. The knowledge of those noises is important in the ECC development to reduce the errors occurrence and therefore improve the channel (i.e., the memory) reliability. The data retention noise is important to be estimated and eventually improved especially for SSD architectures relying on Flash memories since the data recovery process from a retention-corrupted memory cell is a time-consuming and power-hungry task. Nowadays, LDPC are the most popular error correction codes used inside SSDs thanks to their ability of

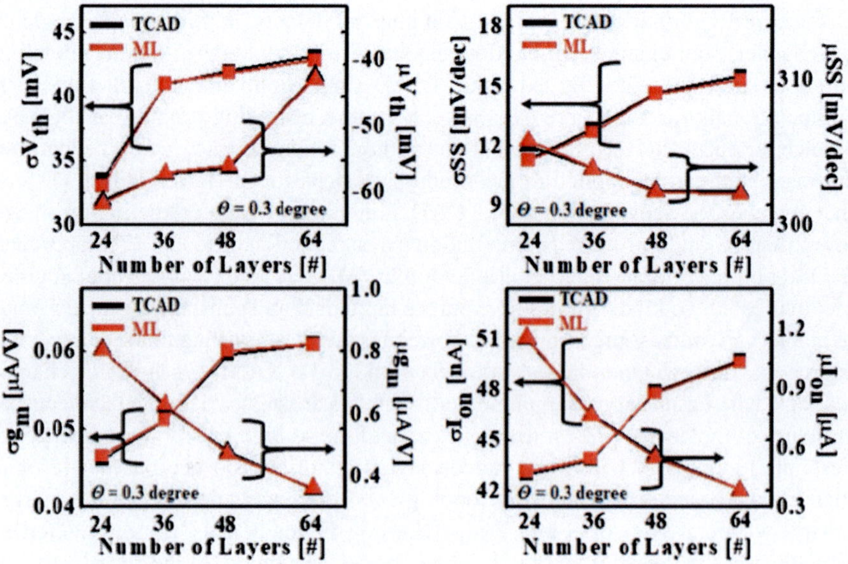

Fig. 6 Variation of the mean and standard deviation of the distribution of the key electrical parameters as the number of stacked layers of 3-D NAND flash memory devices increases. Reproduced with permission from [36]. © 2020 IEEE

handling high NAND raw BERs through soft-decision decoding (SD) at the cost of a longer decoding latency [1]. The decoding performance of the SD operation depends on the accuracy of the Log-Likelihood Ratio (LLR) of the channel coded bits, considering the fact that reads performed at different voltages can provide a better estimation of the LLR values. Let us take as an example a TLC NAND Flash memory storing three bits per cell. To read the data stored in a cell, its threshold voltage is measured and compared against predefined reference (fixed) voltages inside the memory sensing circuit. At least seven read thresholds are needed for a TLC memory. This will generate "hard" outputs of the 3D NAND Flash channel, but to generate the LLR to support SD more read reference voltages should be used. In [38], it is proposed a deep learning approach to dynamically assign the read reference voltages for multi-level Flash memories. Although the work is based on planar technology, it scales very well on the 3D counterpart. With such framework, all the noise source on the memory channel that are unpredictable can be learned from the training data. It is proposed a recurrent neural network (RNN)-based engine to effectively detect the data stored in multi-level cells (from MLC to QLC) without any prior knowledge of the communication channel (i.e., the memory). This is important since it enables the design of powerful ECC engines without a detailed knowledge of the 3D NAND Flash reliability mechanisms. As describer in chapter "Neural Networks and Deep Learning Fundamentals" of this book, the RNN is a class of NNs with feedback connections. It is very suitable to model time series tasks through the memorization of input sequences. In [38], the authors exploit the Gated Recurrent Unit (GRU)

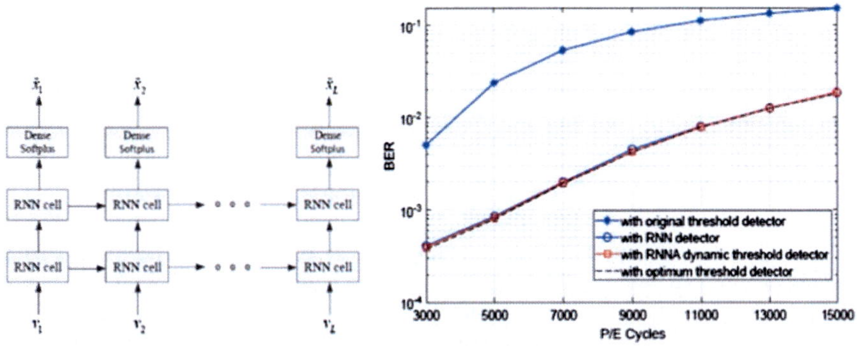

Fig. 7 (Left) RNN architecture with different layers exposed and (right) BER of the RNN detector and RNNA dynamic threshold detector at retention time $T = 10^4$ h for different P/E cycles. Reproduced with permission from [38]. © 2020 IEEE

RNN cell architecture since it employs a lower number of tunable parameters. The network architecture is composed by two GRU layers and a fully connected output layer as shown in Fig. 7. The activation function for GRU layers is the ReLU, whereas for the output layer it is considered a smooth approximation of the ReLU called softplus whose function is based on $\ln(1 + \exp(x))$. The RNN is trained with the readback threshold voltages sensed by the memory cells that are generated by a statistical model developed in [39]. In the simulations, the number of neurons in the input layer is set to 50 and it is found that the optimal training set size considering Adam optimizer and Xavier uniform initializer should be 3×10^6 3D NAND Flash memory cells. For a generic LDPC code whose codeword size is 8000 bits, this translates in 750 recovered codewords for training the RNN. To avoid the longer read latency of an RNN model and the increased power consumption compared with conventional threshold voltage detector applied to ECC, the authors also developed an RNN-aided (RNNA) dynamic threshold detector that can be activated during the idle times of SSDs and that enables the prediction of the SD levels to improve the overall memory reliability. For a TLC memory, 7 read threshold levels $\{q_1, \ldots, q_7\}$ need to be determined. With those discrimination levels we can obtain the hard estimate of the channel while the RNN outputs its proper estimation. Therefore, the adjusted discrimination thresholds $\{q_1{}^*, \ldots, q_7{}^*\}$ can be obtained by searching for the thresholds that can minimize the Hamming distance between the two estimates. This is a compute intensive procedure completely described in [38]. Figure 7 shows the simulated BER performance using the RNN-based detector and the RNNA dynamic threshold at a specific retention time and for different P/E cycles.

4 Machine Learning for Solid State Drives Prognostics

For cloud providers business success is dependent upon the solution of a tough constrained problem, which es requires the hardware resources cost minimization while maximizing their use and reliability in shared or multi-tenant environments. While the solution of such a problem can be found by setting different Service Level Agreement (SLA) terms for computing, memory, and networking devices, it is difficult to achieve the same for storage. The internal architecture of SSDs and the peculiarity of the storage media integrated within, may cause dramatic SLA violations due to either a bad management of the data input/output pipeline or specific issues of the workload submitted to SSD that triggers early reliability failures [40]. For a storage designer it is difficult to directly access the internal architecture of the drive; therefore, going to an higher abstraction level (i.e., algorithms handling the data from/to the drive) is the only viable solution to improve the reliability and the performance of the drive. The set of algorithms applied to these extents are also labeled as prognostics tools. Machine learning emerged as a very promising candidate in the prognostics context because it offers advanced techniques to improve latency, throughput, reliability, and many other SSD's parameters with a minimal system overhead that does not compromise the defined SLA.

4.1 Machine Learning Assisted SSD Lifetime Enhancement

The SSD lifetime extension is achieved primarily by mitigating the endurance stress of the drive caused by repeated data writes and erases. The host-based techniques adopted to reduce the movement of data toward the SSD is known as compression (i.e., reducing the data size by an algorithm that shrinks the original informative content) and deduplication (i.e., an algorithm that reduces that probability to have redundant yet duplicate data) [41]. The effectiveness of those techniques has been proven in many data storage applications bringing an additional benefit related to the increased storage capacity without requiring substantial modifications of the applications and the file system hosted by the SSD. The entire system may even optimize the compression or deduplication process being aware of the storage features, as indicated in [42]. However, using data compress and deduplication still poses some challenges on SSD performance and reliability. Indeed, both operations are not immune from errors (there is a probability that algorithms poorly perform or fail due to hardware or firmware implementation) and require time for calculations and effective data manipulation. The storage, per se, receives few information from the host during the I/O: the data to store and the block address range where data should be saved. In this case, there is no context-related information to avoid the computation overhead on compressing files that are already compressed or use tailored algorithms according to the file typology. One may envision a dedicated hardware accelerator to reduce the computation overhead by offloading the compression or the deduplication on specific engines, but this results in additional hardware integration and increases, in turn, the cost of the SSD. Indeed, we must remind that compression and deduplication are two distinct operations at the system-level, thus translating in a two-step writing process within the storage. To this extent, in [41] it has been presented a machine learning assisted technique for data reduction called Peniseve that can be implemented as a middleware between storage and host. Pensieve is intended as a low-overhead algorithm that is executed directly on the idle CPUs inside the SSD controller. Its architecture is presented in Fig. 8. The interaction with the CPU of the host occurs through a state-of-the-art I/O interface and the Flash Translation Layer is executed by the SSD controller. The components of Pensieve are a classifier that categorizes the incoming data from host into a specific compression class, a set of dictionary files each one belonging to a proper compression class, and the compression/decompression engines assisted by the dictionaries. When the host issues a write command to the SSD through the I/O communication interface, Pensieve looks over a small portion of the dataset and based on that it decides whether it requires a compression or not. This saves computation resources from the SSD controller perspective, relieving the CPUs of this component from such task and giving more time for other background operations like Garbage Collection, Wear Leveling, etc. If the dataset to write must be compressed, then Pensieve decides which compression class and dictionary must adopt and so it starts the operation.

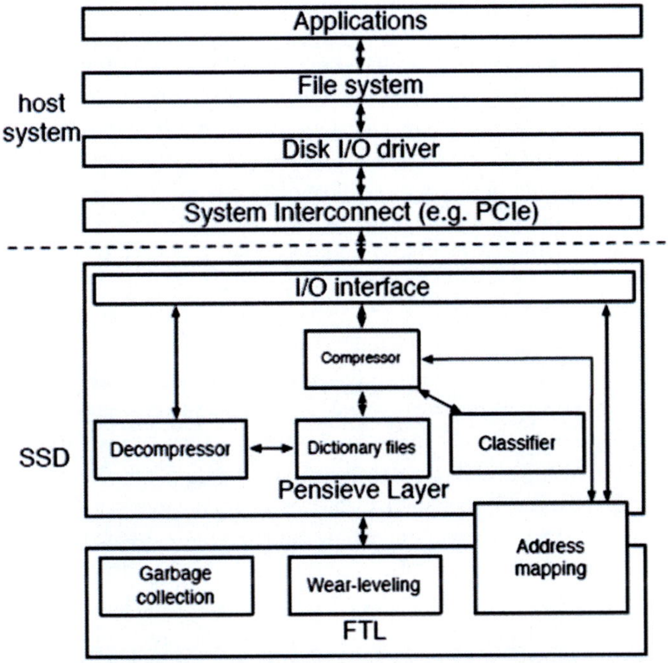

Fig. 8 The system architecture of Pensieve machine learning based compression. Reproduced with permission from [41]. © 2019 IEEE

Since the dictionary information is required in the decompression stage, such tip needs to be included in the metadata associated to the mapping table used to locate the written data correctly. The classifier of Pensieve has been trained on 800,000 files with different sizes and type. The machine learning algorithms evaluated for this classification task ranged from Random Forest, Support Vector Machines, and Boosted Trees. The algorithms were chosen to be easily implementable on general purpose CPUs in modern SSD controllers, while providing high classification accuracy and low computation overhead. The best result was achieved by the Random Forest implementation with 87.16% accuracy and a computation latency of about 13 μs. This result is obtained by setting the amount of data required for prediction down to 512 bytes, so that the compression operation can be started in early phase and the latency of this operation can be hidden through buffering and multitasking, thus exploiting at full the SSD internal architecture. Overall, Pensieve reduces the number of writes to the SSD by 19% with a minimal overhead in the drive's operation latency. Although deduplication is not touched by the work in [41], it is speculated that, while standard approaches keep compression and deduplication separate, the Pensieve's approach inherently achieves some sort of data deduplication through compression, even though not with the same results of a dedicated algorithm.

4.2 *Echo State Networks for Hot Data Prediction in SSDs*

With both 3D and planar NAND Flash technologies it is not possible to do an in-place update operation; therefore, each time we want to modify the stored data, we can only proceed with a new write operation. This comes at a price of a reduced lifespan of the memory module and therefore of the SSD integrating these devices. The Flash Translation Layer (FTL) running inside SSD controllers is exaclty designed to take care of the above-mentioned issue, by means of two firmware modules: Garbage Collection (GC) and Wear Levelling (WL) [43]. The former is responsible for reclaiming the storage space occupied by invalid or old data to accommodate potentially updated or new data; the latter sends frequently written data (a.k.a. hot data) to blocks that experienced a low endurance stress (i.e., less program/erase cycles) while it dispatches least recently accessed data (a.k.a. the cold data) to blocks that already sustained higher wear-out (i.e., more program/erase cycles). Both GC and WL must work together to produce the best outcome at the drive level. With the above considerations in mind, the identification of the blocks storing hot and cold data is mandatory to extend the endurance of the NAND Flash and, therefore, of the SSD itself. The Hot Data Identification (HDI) process uses a large statistical characterization of the workload (i.e., the application and the characteristics file system mounted on the drive) submitted to the SSD [44]. The literature provides examples related to HDI procedures based on the SSD access behavior characterization and on stochastic models [45–47]. However, it is found that the identification of the hot blocks in a workload is a time-variant task, complicating those algorithms profoundly. Motivated by this issue, in [48] it has been proposed for the first time the use of an innovative machine learning tool based on

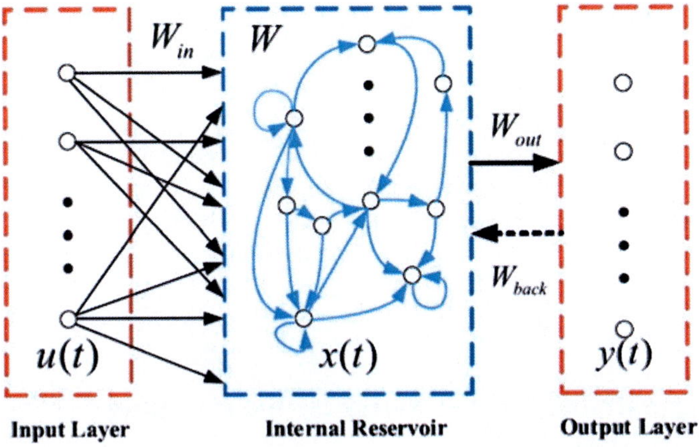

Fig. 9 The structure of an Echo State Network. Reproduced with permission from [48]. © 2020 IEEE

the reservoir computing concept, namely the Echo State Network (ESN). An ESN, as shown in Fig. 9, is a neural network including an input layer, a hidden layer made by recurrent connected neurons (this is also called the dynamic reservoir), and a connected output layer. As highlighted by the figure, the input layer is connected to the reservoir layer via the weights W_{in}. The dynamic reservoir has internal weights W and is connected to the output layer through output weights W_{out}. Then there is a feedback structure of the output in reservoir through feedback weights W_{back}. The property that makes the difference between an ESN and a RNN approach lies exactly on the reservoir concept. In RNN, the training is achieved by minimizing the mean square error of the model during the learning of the input and of the output weights. The ESN needs to learn only the weights of the neurons in the reservoir. Using a neural network for the hot block prediction task turns the HDI problem into a HDP (Hot Data Prediction) one, avoiding then the issues related to a prior characterization of the workload sustained by the SSD and achieving real-time hot/cold data separation. It is worth to point out that the method of [48] is orthogonal to state-of-the-art HDI algorithms already present in most of the Flash Translation Layer firmware, since those HDI systems can be augmented by the information mined by the HDP based on the ESN. The methodology developed in [48] has been tested on an SSD prototype developed in a lab environment that has been stressed with different workloads. Among them, we point out an On-Line Transaction Processing (OLTP) either write-intensive or read-intensive, and a write-intensive workload mimicking an industrial environment. The advantages shown by the ESN-HDP are many. A reduced energy consumption by 3.4% is achieved with respect to HDI algorithms since less write operations are triggered on the NAND Flash memories integrated in the SSD. The number of block erase operations triggered by the GC is reduced by almost 1,000 times since less data updates operations are to be invoked. The average response time of the drive, critical parameter in SSD especially for the Quality-of-Service (QoS) concept, is reduced by 10 to 30 μs. The disadvantages of this kind of approach are mainly on the complexity of the training procedure, which requires an appropriate smart methodology named Swarm Particle Optimization (SPO) to be implemented by the user to find the best reservoir characteristics (e.g., number of neurons, scaling coefficient, etc.) just to avoid the uncertainty and the inconvenience of setting those parameters by hand. Another factor to account for is the RAM occupation when the ESN is deployed in an SSD controller, although it has been shown by the authors in [48] that this overhead can be comparable with other HDI approaches presented in literature.

5 Computational Storage and Machine Learning

The computing and the storage requirements of conventional server architectures that populate today's exa-scale computing domain are rapidly hitting the performance limits of CPUs, GPUs, and SSDs. A major issue that system developers are trying to solve is related to the drop of that metric due to the perceived incommunicability of the computing and the storage world. In fact, the storage comes into play only when the computation already took place or just before the start of some calculations, so the data needs to bounce frequently from SSDs to CPUs and GPUs with evident drawbacks on performance, power consumption and QoS. To be said, the computation does not take place where data actually belong (i.e., the storage). Data-intensive applications like high performance computing (HPC), machine learning and artificial intelligence (ML & AI), online databases and many others suffer from this sort of domain separation between computing and storage. To this extent, the computational storage (CS) paradigm has been proposed to the research community [49]. The CS seeks not only an off-load of some computing tasks directly in the storage, but also an overall performance acceleration, reduced power consumption, and reliability control at different levels. It is worth to mention that CS can be achieved with multiple architectures and paradigms that could involve the single 3D NAND Flash device or the entire SSD element [50]. What is required for its implementation is the enabling element that interconnects through a fabric the computing and the storage elements, that in enterprise environments is generally represented by the PCIe protocol [51], as shown in Fig. 10 where a CS architecture is proposed from [52]. With the CS approach it is possible to bring machine learning directly in the storage element with benefits that could never be possible with the traditional infrastructures.

5.1 The Computational Approximate Storage Concept

As said in the introduction of this section, in some machine learning (ML) applications like those for image or speech recognition, it would be beneficial to leverage on a CS environment. In [53], it is proposed a slight variation of the CS by introducing

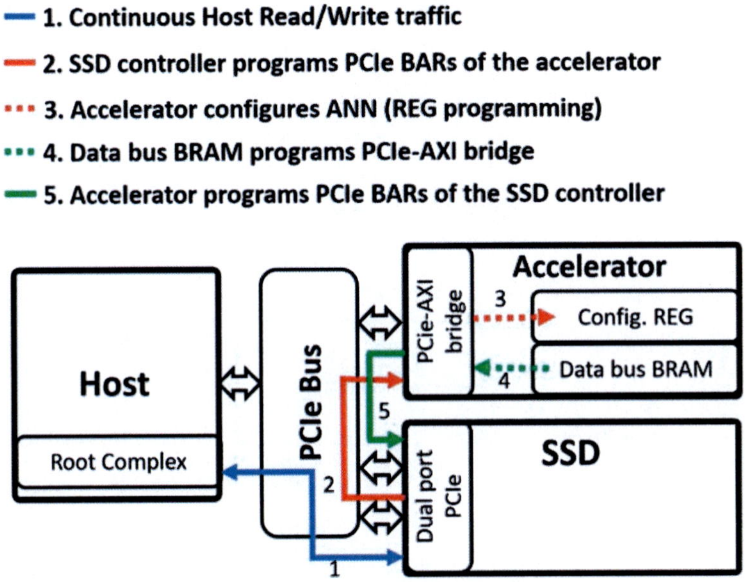

Fig. 10 Schematic of the data transfers between an SSD and a neural network computing accelerator in a CS environment. Reproduced with permission from [52]. © 2019 IEEE

the Computational Approximate Storage (CAS) concept shown in Fig. 11. The CAS, similarly, to CS, proposes to offload the computation of the data manipulation tasks like multiplications and accumulate (i.e., the principal operations performed in the machine learning applications) from CPUs and GPUs directly to storage, thus minimizing the data bounce. The innovation here is to exploit the resiliency to errors of the Machine Learning algorithms by introducing the approximate computing concept [54]. Being tolerant with respect to the memory errors exposed by the 3D NAND Flash technology, the core of the storage element (i.e., SSD), power consumption and performance metrics are improved. However, if machine learning recognition tasks are performed directly on memory, the number of errors can be excessive in some working conditions and drastically impact the outcome of the process. Neural networks indeed have a BER limit of the inputs under which it is impossible to make them work to a reasonable extent. Knowing this, to keep the memory errors under control and thus enable the CAS paradigm, it is proposed a Neural Network-based Memory Error Patrol (MEP) directly applied to the 3D-TLC NAND flash memories in the storage. Through a prediction of the error patterns and of the threshold voltage shifts in the memory cells due to endurance, retention, or read disturb stress, it is possible to choose which error reduction operation (e.g., read reference shift, soft decoding, etc.) is best for reducing the errors amount. The MEP engine in [53] is implemented in the memory controller which is stacked over the Flash memory modules to save integration area and increase computational speed. The role of the MEP is split into two different units: a State Shift Error Prediction (SSEP) unit and

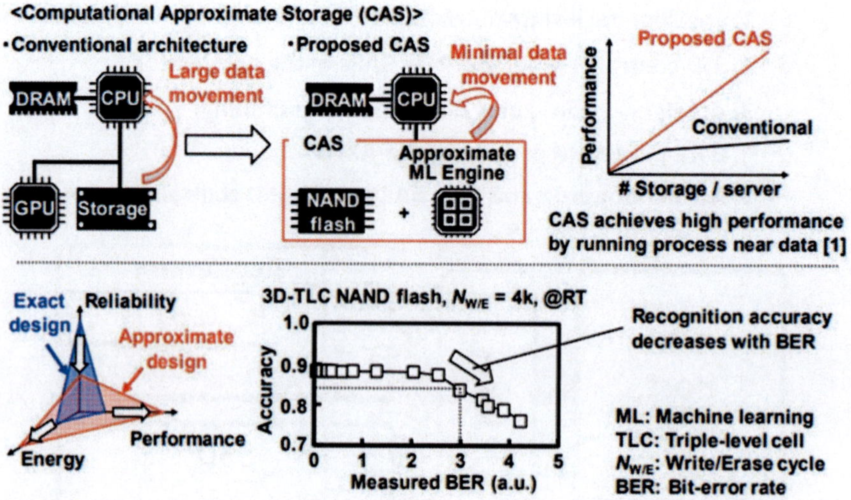

Fig. 11 Concept of proposed computational approximate storage (CAS). Error patrol is needed to achieve high performance by acceptable reliability loss. Reproduced with permission from [53]. © 2020 IEEE

an Error Data Pattern Prediction (EDPP) unit. The former unit is responsible for the prediction of the error's location and magnitude. Prediction is achieved through three distinct neural networks acting on different dimensions (i.e., memory Block, single page, or Flash string). The training of the SSEP is intended to be performed during the 3D NAND Flash qualification test flow, which follows the manufacturing process; it takes about 50 Mbytes of training data related to the threshold voltage shift of the cells under test, measured under different conditions of endurance stress, retention, and so on. Remarkably good results are obtained by predicting the memory BER with 2.6% errors even when affected by inter-chip variations. The EDPP unit is on the contrary a simpler unit. It is made by a single neural network that guesses the physical origin of the errors including the information for neighbor pages. By combining both SSEP and EDPP, the 3D NAND Flash error rate can be successfully monitored, and the computational approximate storage is realized.

5.2 A Neural Network Engine Example in Computational Storage

Inside SSD drives, there is a complex system of algorithms devoted to monitor and manage the reliability of the storage medium; in essence, a very tight collaboration between Flash memories and the brain of the storage device, namely the SSD controller. Throughout the whole SSD lifetime, especially during idle times, here is a set of operations that the controller performs to guarantee a high intrinsic reliability level by tuning the 3D NAND Flash memories read reference voltages which, in the end, are required to retrieve the stored information from the SSD. The above-mentioned voltage tuning is also known as read-retry or V_T-shift [1]. Up to now, this algorithm has been implemented in the SSD controller firmware by a set of dedicated look-up tables stored in the controller memory that associates the optimal read retry operation or V_T-shift to each read reference voltage of each 3D NAND Flash (or a portion of thereof) at a peculiar SSD lifetime point. With the increased complexity of 3D Flash memories this approach is starting to run out of gas. The idea presented in [52] is to off-load the execution of this voltage tuning from the SSD controller to a Field-Programmable Gate Array (FPGA)-based neural network accelerator that embraces the computational storage paradigm. This can be achieved through direct communication between storage and accelerator computing elements using the PCIe interconnection fabric, as shown in Fig. 12. The key contribution of [52] is a design of a combined feed-forward regression artificial neural network (ANN) and of a non-linear auto-regressive with exogenous input (NARX) ANN that predict the optimal read reference voltages to maximize the 3D NAND Flash lifetime, and thus the SSD reliability/performance. The regression ANN is a single hidden layer network whose size is $8 \times 10 \times 1$. The activation function is a symmetric saturating linear transfer function for both hidden and output layers. The eight input neurons represent topological information of the NAND Flash and the read reference voltages at the beginning of the memory lifetime. The NARX ANN has the only function to predict, through a time-series analysis of the temperature readings of the SSD, what will be the temperature of the drive in the upcoming data readout and this information is used as an input to the regression ANN. The prototype of the accelerator has been

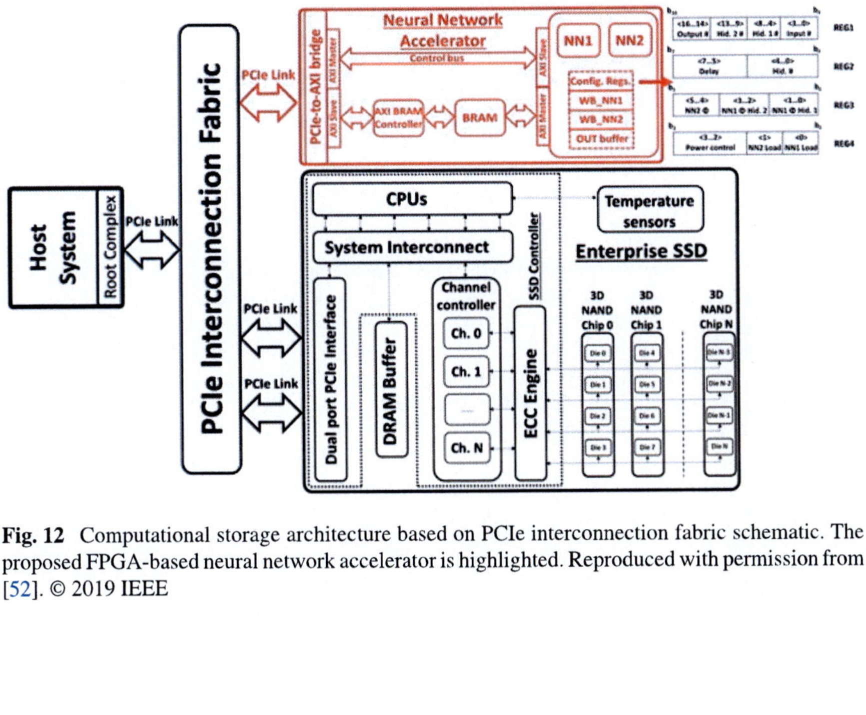

Fig. 12 Computational storage architecture based on PCIe interconnection fabric schematic. The proposed FPGA-based neural network accelerator is highlighted. Reproduced with permission from [52]. © 2019 IEEE

developed on a Xilinx VC707 FPGA demonstrating a very high prediction accuracy (up to 99.5%) of the read reference voltages under different working conditions and a low resource utilization (less than 10% of the FPGA) making room for the implementation of additional algorithms or accelerators based on the CS paradigm.

6 Conclusions

In this chapter, we provided an overview of the machine learning techniques explored in improving the reliability and the performance of Solid State Drives based on the 3D NAND Flash storage medium. State of the art supervised and unsupervised techniques like clustering are applied for Error Correction Codes optimization to improve the drive resilience against endurance and retention errors typical of the Flash technology. Advanced approaches like deep learning methodologies based on different neural network architectures are exploited in coping with large reliability and performance variability issues. Finally, the use of these techniques is shown in the context of the Computational Storage Paradigm to prove the large benefits in adopting such architectures. All the presented methodologies are based on training and testing procedures that require large datasets populated with device and system-level characterizations performed in many usage corners both of the storage media and of the drive.

References

1. L. Zuolo et al., Solid-state drives: memory driven design methodologies for optimal performance. Proc. IEEE **105**(9), 1589–1608 (2017)
2. R. Micheloni et al., Array architectures for 3-D NAND flash memories. Proc. IEEE **105**(9), 1634–1649 (2017)
3. Y. Fukuzumi et al., Optimal integration and characteristics of vertical array devices for ultra-high density, bit-cost scalable flash memory, in *IEEE International Electron Devices Meeting (IEDM)*, Dec 2007, pp. 449–452
4. M. Ishiduki et al., Optimal device structure for pipe-shaped BiCS flash memory for ultra high density storage device with excellent performance and reliability, in *IEEE International Electron Devices Meeting (IEDM)*, Dec 2009, pp. 1–4
5. W. Jeong et al., A 128 Gb 3b/cell V-NAND flash memory with 1 Gb/s I/O rate. IEEE J. Solid-State Circuits **51**(1), 204–212 (2016)
6. T. Tanaka et al., A 768 Gb 3b/cell 3D-floating-gate nand flash memory, in *IEEE International Solid-State Circuits Conference (ISSCC)*, Jan 2016, pp. 142–144
7. H. Kim et al., Evolution of NAND flash memory: from 2D to 3D as a storage market leader, in *IEEE International Memory Workshop (IMW)*, May 2017, pp. 1–4
8. N. Righetti et al., 2D vs 3D NAND technology: reliability benchmark, in *IEEE International Integrated Reliability Workshop (IIRW)*, Oct 2017, pp. 1–6
9. C. Zambelli et al., Cross-temperature effects of program and read operations in 2D and 3D NAND flash memories, in *IEEE International Integrated Reliability Workshop (IIRW)*, Oct 2018, pp. 1–4

10. C. Zambelli et al., First evidence of temporary read errors in TLC 3D-NAND flash memories exiting from an idle state. IEEE J. Electron Devices Soc. **8**, 99–104 (2020)
11. B. Shin et al., Error control coding and signal processing for flash memories, in *IEEE International Symposium on Circuits and Systems (ISCAS)*, May 2012, pp. 409–412
12. C. Zambelli et al., Reliability challenges in 3D NAND flash memories, in *IEEE International Memory Workshop (IMW)*, May 2019, pp. 1–4
13. H. Maejima et al., A 512Gb 3b/cell 3D flash memory on a 96-word-line-layer technology, in *IEEE International Solid State Circuits Conference (ISSCC)*, Feb 2018, pp. 336–338
14. S. Lee et al., A 1Tb 4b/cell 64-stacked-WL 3D NAND flash memory with 12 MB/s program throughput, in *IEEE International Solid State Circuits Conference (ISSCC)*, Feb 2018, pp. 340–342
15. N.R. Mielke et al., Reliability of solid-state drives based on NAND flash memory. Proc. IEEE **105**(9), 1725–1750 (2017)
16. L. Zuolo et al., LDPC soft decoding with reduced power and latency in 1X-2X NAND flash-based solid state drives, in *IEEE International Memory Workshop (IMW)*, May 2015, pp. 1–4
17. J. Yang, High-efficiency SSD for reliable data storage systems, in *Proceedings of the Flash Memory Summit* (2012)
18. D. Hogan et al., Estimating MLC NAND flash endurance: a genetic programming based symbolic regression application, in *Proceedings of the Conference on Genetic and Evolutionary Computation* (2013), pp. 1285–1292
19. T. Arbuckle et al., Learning predictors for flash memory endurance: a comparative study of alternative classification methods. Int. J. Comput. Intell. Stud. **3**(1), 18–39 (2014)
20. Y. Nakamura et al., Machine learning-based proactive data retention error screening in 1Xnm TLC NAND flash, in *International Reliability Physics Symposium (IRPS)*, Apr 2016, pp. PR–3–1–PR–3–4
21. C. Zambelli et al., Characterization of TLC 3D-NAND flash endurance through machine learning for LDPC code rate optimization, in *IEEE International Memory Workshop (IMW)*, May 2017, pp. 1–4
22. A. Marelli, R. Micheloni, *BCH and LDPC Error Correction Codes for NAND Flash Memories* (Springer, Netherlands, 2016), pp. 281–320
23. X. Hu, LDPC codes for flash channels, in *Proceedings of the Flash Memory Summit* (2012)
24. J. MacQueen, Some methods for classification and analysis of multivariate observations, in *Proceedings of Berkeley Symposium on Mathematical Statistics and Probability*
25. B. Fitzgerald et al., Endurance prediction and error reduction in NAND flash using machine learning, in *Non-Volatile Memory Technology Symposium (NVMTS)* (2017), pp. 1–8
26. K.R. Muller et al., An introduction to kernel-based learning algorithms. Trans. Neur. Netw. **12**(2), 181–201 (2001)
27. E. Gal et al., Algorithms and data structures for flash memories. ACM Comput. Surv. (2005)
28. Y. Nakamura et al., Machine learning-based proactive data retention error screening in 1Xnm TLC NAND flash, in *IEEE International Reliability Physics Symposium (IRPS)* (2016), pp. PR-3-1–PR-3-4
29. O. Arbelaitz et al., Applying resampling methods for imbalanced datasets to not so imbalanced datasets, in *Advances in Artificial Intelligence* (2013), pp. 111–120
30. T. Nakamura et al., Adaptive artificial neural network-coupled LDPC ECC as universal solution for 3-D and 2-D, charge-trap and floating-gate NAND flash memories. IEEE J. Solid-State Circuits **54**(3), 745–754 (2019)
31. T. Tokutomi et al., Advanced error prediction LDPC for high-speed reliable TLC nand-based SSDs, in *IEEE International Memory Workshop (IMW)*, May 2014, pp. 1–4
32. A. Spinelli et al., Reliability of NAND flash memories: planar cells and emerging issues in 3D devices. Computers **6**(2), 16 (2017)
33. Y. Yanagihara et al., Control gate length, spacing and stacked layer number design for 3D-stackable NAND flash memory, in *IEEE International Memory Workshop*, May 2012, pp. 1–4
34. K.T. Kim et al., The effects of taper-angle on the electrical characteristics of vertical NAND flash memories. IEEE Electron Device Lett. **38**(10), 1375–1378 (2017)

35. Y.-T. Oh et al., Impact of etch angles on cell characteristics in 3D NAND flash memory. Microelectron. J. **79**, 1–6 (2018)
36. K. Ko et al., Variability-aware machine learning strategy for 3-D NAND flash memories. IEEE Trans. Electron Devices **67**(4), 1575–1580 (2020)
37. D.C. Lee et al., Machine learning model for predicting threshold voltage by taper angle variation and word line position in 3D NAND flash memory. IEICE Electron. Express **17**(22), 20200345 (2020)
38. Z. Mei et al., Deep learning-aided dynamic read thresholds design for multi-level-cell flash memories. IEEE Trans. Commun. **68**(5), 2850–2862 (2020)
39. G. Dong et al., Enabling nand flash memory use soft-decision error correction codes at minimal read latency overhead. IEEE Tran. Circuits Syst. I Regul. Pap. **60**(9), 2412–2421 (2013)
40. J.-E. Dartois et al., Investigating machine learning algorithms for modeling SSD I/O performance for container-based virtualization. IEEE Trans. Cloud Comput. 1–14 (2019)
41. I. Te, M. Lokhandwala et al., Pensieve: a machine learning assisted SSD layer for extending the lifetime, in *IEEE International Conference on Computer Design (ICCD)* (2018), pp. 35–42
42. A. Zuck et al., Compression and SSDs: where and how?, in *2nd Workshop on Interactions of NVM/Flash with Operating Systems and Workloads*, USENIX Association (2014)
43. M. Yang et al., Garbage collection and wear leveling for flash memory: past and future, in *International Conference on Smart Computing* (2014), pp. 66–73
44. D. Park et al., Hot data identification for flash-based storage systems using multiple bloom filters, in *Proceedings of the IEEE MSST*, May 2011, pp. 1–11
45. M.W. Lin et al., HDC: an adaptive buffer replacement algorithm for NAND flash memory-based databases. Optik **125**(3), 1167–1173 (2014)
46. J.-W. Hsieh et al., Efficient identification of hot data for flash memory storage systems. ACM Trans. Storage **2**(1), 22–40 (2006)
47. J. Liu et al., A novel hot data identification mechanism for NAND flash memory. IEEE Trans. Consum. Electron. **61**(4), 463–469 (2015)
48. Q. Luo et al., Self-learning hot data prediction: where echo state network meets NAND flash memories. IEEE Trans. Circuits Syst. I Regul. Pap. **67**(3), 939–950 (2020)
49. SNIA Computational Storage Technical Work Group (TWG), https://www.snia.org/computational (2019)
50. P. Mehra, Samsung SmartSSD: accelerating data-rich applications, in *Presented at Flash Memory Summit*, Aug 2019
51. PCI Express Base 3.0 Protocol, http://www.pcisig.com/ (2019)
52. C. Zambelli et al., Enabling computational storage through FPGA neural network accelerator for enterprise SSD. IEEE Trans. Circuits Syst. II Express Briefs **66**(10), 1738–1742 (2019)
53. M. Abe et al., Computational approximate storage with neural network-based error patrol of 3D-TLC NAND flash memory for machine learning applications, in *IEEE International Memory Workshop (IMW)* (2020), pp. 1–4
54. C. Matsui et al., Application-induced cell reliability variability-aware approximate computing in TaOX-based ReRAM data center storage for machine learning, in *IEEE Symposium VLSI Technology*, June 2019, pp. T234–T235

Index

R. Micheloni and C. Zambelli (eds.), *Machine Learning and Non-volatile Memories*, https://doi.org/10.1007/978-3-031-03841-9